THRESHOLD

Cambridge Pre-GED Program in Mathematics

1

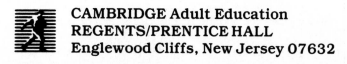
CAMBRIDGE Adult Education
REGENTS/PRENTICE HALL
Englewood Cliffs, New Jersey 07632

Library of Congress Cataloging-in-Publication Data

Threshold : Cambridge pre-GED program in mathematics 1.
p.   cm.
ISBN 0–13–110966–9 (pbk.)
1. Mathematics—Problems, exercises, etc.   2. General educational
development tests—Study guides. I. Title.
QA139.T47   1992
513′.076—dc20   91–23917
                                                                  CIP

Publisher: TINA B. CARVER
Executive Editor: JAMES W. BROWN
Editorial Supervisor: TIMOTHY A. FOOTE
Managing Editor: SYLVIA MOORE
Production Editor: JANET S. JOHNSTON
Copyeditor: TECHNICAL EDITING SERVICES, INC.
Pre-press Buyer: RAY KEATING
Manufacturing Buyer: LORI BULWIN
Scheduler: LESLIE COWARD
Interior designers: JANET SCHMID and JANET S. JOHNSTON
Artists: JANET S. JOHNSTON, WARREN FISCHBACH,
and CAROL HYLAND
Cover coordinator: MARIANNE FRASCO
Cover designer: BRUCE KENSELAAR
Cover photo: J. LANGE/ STOCK IMAGERY, INC.
Photo researcher: JULIE SCARDIGLIA
Permissions: ELLEN DIAMOND

 © 1993 by REGENTS/PRENTICE HALL
A Division of Simon & Schuster
Englewood Cliffs, New Jersey 07632

Printed in the United States of America

10   9   8   7   6   5   4   3   2   1

ISBN 0-13-110966-9

PHOTO CREDITS
page 7   Hakim Raquib
125   Laimute E. Druskis
199   Teri Leigh Stratford

Prentice-Hall International (UK) Limited, *London*
Prentice-Hall of Australia Pty. Limited, *Sydney*
Prentice-Hall Canada Inc., *Toronto*
Prentice-Hall Hispanoamericana, S.A., *Mexico*
Prentice-Hall of India Private Limited, *New Delhi*
Prentice-Hall of Japan, Inc., *Tokyo*
Simon & Schuster Asia Pte. Ltd., *Singapore*
Editora Prentice-Hall do Brasil, Ltda., *Rio de Janeiro*

CONTENTS

ACKNOWLEDGMENTS

CAMBRIDGE Adult Education thanks the men and women enrolled in ABE and Pre-GED courses who read parts of the *Threshold* manuscripts and offered valuable advice to the programs' authors and editors.

We also thank the following consultants for their many contributions throughout the preparation of the *Threshold* Pre-GED programs.

Cecily Kramer Bodnar
Consultant, Adult Learning
Adult Literacy Services
Central School District
Greece, New York

Pamela S. Buchanan
Instructor
Blue Ridge Job Corps Center
Marion, Virginia

Maureen Considine, M.A., M.S.

Learning Laboratory Supervisor
Great Neck Adult Learning Center
Great Neck, New York

ABE/HSE Projects Coordinator
National Center for Disability Services
Albertson, New York

Carole Deletiner
Instructor
Hunter College
New York, New York

Patricia Giglio
Remedial Reading Teacher
Johnstown ASACTC
Johnstown, New York

Diane Marinelli Hardison, M.S. Ed.
Mathematics Educator
San Diego, California

Margaret Banker Tinzmann, Ph. D.
Program Associate
The North Central Regional Educational Laboratory
Oak Brook, Illinois

The *Threshold* Pre-GED Programs

Threshold provides a full-range entry-level course for adults whose goal is to earn a high school equivalency diploma. The men and women who use the six *Threshold* programs will learn—and profit from an abundance of sound practice in applying—the writing, problem-solving, and critical-reading and -thinking skills they'll need when they take the GED tests. They will gain a firm grounding in knowledge about social studies and science and will read many excellent selections from the best of classical and contemporary literature. In short, *Threshold* offers adults the skills, knowledge, and practice that will enable them to approach GED-level test preparation with well-deserved confidence and solid ability.

The *Threshold* Mathematics Program

Two books make up the mathematics program. This volume covers whole numbers and decimals, and the second covers fractions, percents, and tables and graphs.

Students should begin their study in this book by taking the Pretest. It has two parts, which can be administered at the same time or separately. To aid accurate assessment, the Pretest's problems are not multiple choice. To facilitate student placement, the test has at least one problem related to each lesson in this book, as the Skills Chart that follows it shows.

The lessons present the various facets of whole number and decimal operations and applications in a carefully graded skill-building sequence. Lessons are typically divided into subskill segments, each with a succinct explanation, an example worked out and explained step by step, an exercise, and a word problem. A glance at Lesson 5 illustrates these points. After the lesson gives the basic addition facts, it covers the "no-borrowing" subskills of whole-number addition in seven distinct segments, each building on the previous.

In each unit Mixed Practices follow the lessons on subtraction and division. A comprehensive review of all four operations precedes the last chapter in each unit.

Both units' final chapters cover applications and one-step and multi-step problem solving. The lessons show how to use the four basic operations—sometimes in simple formulas—to solve the kinds of averaging, cost, measurement, and basic geometry problems adults encounter at work, in their everyday lives, and on the GED.

Each unit ends with a thorough review followed by a GED Practice. The Practices, formatted like the GED, provide further review and valuable test-taking experience. Accompanying skills charts allow assessment of skill mastery.

The Posttest, also formatted like the GED, has at least one problem for each of the lessons in the units' final chapters.

This book's quick and accurate placement tool, its carefully segmented instruction with word problems throughout, its frequent cumulative exercises, its chapters on problem solving and applications, its progress-assessment charts, and its GED-like practices and Posttest make it—together with the second volume—an excellent first course in preparation for the mathematics test of the GED.

TO THE STUDENT

You will profit in several important ways by using this book as you begin to prepare for the mathematics test of the GED:

- You will improve your math skills.
- You will increase your problem-solving ability.
- You will gain experience in answering questions like those on the GED.
- You will become more confident of your abilities.

To Find Out About Your Current Math Skills. . .

Take the PRETEST. When you have finished, refer to the ANSWERS AND SOLUTIONS at the back of this book to check your answers. Then look at the SKILLS CHART that follows the Pretest. It will give you an idea about which parts of this book you need to concentrate on most.

To Improve Your Math Skills and Problem-Solving Ability . . .

Study the LESSONS. They present instruction about various math skills and about the methods for solving problems. Each lesson includes EXAMPLES and one or more EXERCISES to help you improve both your math skills and your ability to solve problems.

Do the problems in the two MIXED PRACTICES and the two REVIEWS in each unit. The problems they contain offer you further practice and provide a way for you to review.

To Gain Experience in Solving Problems Like Those on the GED . . .

Take the GED PRACTICE at the end of each unit. The GED Practices are made up of problems like the ones on the mathematics test of the GED. They offer test-taking experience that you will find useful when you take the GED.

Before you finish with this book, take the POSTTEST. Like the two GED Practices, it is similar to the GED's mathematics test. Look at the SKILLS CHART that follows the Posttest. If you compare your Pretest and Posttest performances, you will probably find that your math skills and problem-solving ability have improved as you have worked through this book. The chart can give you an idea about which parts of this book you should review.

Pretest

If you begin this book at the very beginning, you may find that you already know how to do some of the work. This pretest can help you decide if you really need to start at page 8, or if you can skip some lessons. Answer as many of the questions as you can and then check your answers. The Skills Chart on page 4 will help you find where you should start your work in this book.

MATHEMATICS PRETEST

Directions: Answer as many questions as you can.

Whole Number Problems

1. What is the value of the 5 in the number 4750?

2. Write 14,065 in words.

3. Write one hundred six thousand, seventy-five in numbers.

4. Write these numbers in order from smallest to largest:
101, 1001, 1110, 1011, 111, 100, 1010, 11

5. Round 12,095 to the nearest hundred.

6.   4
   + 5

7.   14
   + 23

8.   235
   + 143

9.   13
      11
      24
   + 41

10.   257
   + 365

11.   349
      715
      121
   + 212

12. Add:
23 + 6709 + 10 + 302 + 405 =

13. Find the difference between 7 and 97.

14.   8
   − 5

15.   47
   − 23

16.   4952
   − 3421

17. 7295 − 6167 =

18. Find the answer to seven times six.

19.   5
   × 4

20.   43
   × 2

21.   2030
   × 458

22.   549
   × 130

23. 3⟌9

24. 6 ÷ 3 =

25. 6⟌306

26. 458⟌929,740

27. Complete the following:
    (a) 1 quart = ____ pints
    (b) 32 ounces = ____ pounds
    (c) 1 yard = ____ feet
    (d) 5280 feet = ____ mile(s)
    (e) 1 year = ____ days
    (f) 1 ton = ____ pounds

28. In the metric system, length or distance is measured in
    (1) feet
    (2) grams
    (3) meters

29. Jorge wants to build a set of shelves for his son's toys. If he has a board 8 feet long, how many 32-inch shelves can he cut from it?

30. Kurt needs to put a fence around his yard to keep his puppy safe at home. His yard is 40 feet wide and 30 feet long.
    (a) How much fencing will he need?
    (b) The vet told Kurt that his puppy should have at least 80 square feet to run in. Will the fence allow enough room?

31. The high temperatures one week were as follows: 87, 89, 91, 88, 86, 92, and 90. What was the average high temperature that week?

32. Fresh shrimp are on sale this week, and Maria wants to buy some for a special treat. If one pound of shrimp costs $6, how much will it cost to buy 24 ounces of shrimp?

Decimal Problems

33. Which digit is in the thousandths place in 3296.0184?

34. Write .52 in words.

35. Write these numbers in order from smallest to largest:
    1.01, 1.001, 1.11, 10.11, 11.1, .0011, 1.0101, .11

36. Round 2.864 to the nearest tenth.

37. $5.25 + $4 =

38. 4509 + 72.2 + 6 + 5002 =

39. 8306.95 − 7317.80 =

40. $\begin{array}{r} 3.9 \\ \times\ 3 \\ \hline \end{array}$

41. $\begin{array}{r} 1.09 \\ \times\ 3.2 \\ \hline \end{array}$

42. $5\overline{)5.5}$

43. $1.2\overline{)40.8}$

44. $.5\overline{)254,609}$

45. Sandra earns $5.50 per hour at her new job. If she works 6 hours on Monday, 7 hours on Tuesday, 7 hours on Wednesday, 8 hours on Thursday, and 7 hours on Friday, how much will she earn that week?

46. Than's doctor wants her to eat no more than 1 gram of salt in a day. How many milligrams of salt is that?

47. The rainfall the first five days of a week was recorded as follows: .87 in., .5 in., .73 in., 1.01 in., and 2 in. What was the average daily rainfall during that period?

48. What is the area of this triangle?

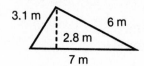

49. Find the circumference and the area of this circle. Round the answers to the nearest tenth of a centimeter.

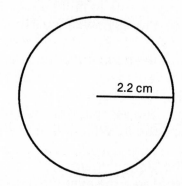

50. Which can of tuna costs less per ounce?
    (1) 6 ounces for $.89
    (2) 12 ounces for $1.75

Check your answers on page 205.

MATHEMATICS PRETEST SKILLS CHART

This chart shows which lesson covers the mathematics skills tested by each item in the Pretest. You can study as much of this book as you wish, but you should study at least those lessons related to any items you missed on the test.

Unit 1	Whole Numbers	Item Number
Lesson 1	Place Value	1
Lesson 2	Reading and Writing Whole Numbers	2, 3
Lesson 3	Ordering Whole Numbers	4
Lesson 4	Rounding Whole Numbers	5
Lesson 5	Adding Whole Numbers	6, 7, 8, 9
Lesson 6	Carrying in Addition	10, 11, 12
Lesson 7	Subtracting Whole Numbers	13, 14, 15, 16
Lesson 8	Borrowing in Subtraction	17
Lesson 9	Multiplying Whole Numbers	18, 19, 20, 21

UNIT 1

Whole Numbers

Chapter 1 of this unit introduces whole numbers. In Chapters 2–5, you will practice the four operations—addition, subtraction, multiplication, and division—using whole numbers. In Chapter 6 you will use whole numbers to solve everyday problems—problems like many of those on the GED. You will work with measurements, perimeter and area, averages, and one-step and multistep word problems.

Unit 1 Overview

Chapter 1 Understanding Whole Numbers
Chapter 2 Addition
Chapter 3 Subtraction
Chapter 4 Multiplication
Chapter 5 Division
Chapter 6 Using Whole Numbers

GED Practice 1

1 UNDERSTANDING WHOLE NUMBERS

The lessons in this chapter introduce whole numbers. They cover place value, reading and writing whole numbers, ordering whole numbers, and rounding whole numbers.

Lesson 1

Place Value

Think about how important numbers are to you. They tell you how much money you earn—and how much you spend. They help you total up the hours you work each week. They let you know if you have enough money to buy a new car. They even announce your birthday. A world without numbers would be a confusing place indeed.

Digits

Whole numbers are written in digits. From the smallest to the largest, the digits are 0, 1, 2, 3, 4, 5, 6, 7, 8, and 9. Using only these digits, you can write any whole number. An example of a one-digit number is 6. A two-digit number is 25. A three-digit number is 789.

Ones, Tens, and Hundreds Places

Every place, or position, in a whole number has a certain value, or worth. Here are the first three whole-number places and their values.

	hundreds	tens	ones
Place Names	hundreds	tens	ones
Places	——	——	——
Place Values	100	10	1

The names of the places in the chart are, from right to left, ones, tens, and hundreds. As you move left, the value of each place is 10 times bigger.

- The ones place has the value of 1.
- The tens place has the value of 10 ones, or 10.
- The hundreds place has the value of 10 tens, or 100.

The value of a digit in a number depends on its position in the number. For example, the 6 in 600 has a larger value than the 6 in 60. (Wouldn't you rather earn $600 a week than $60 a week?)

Think of the value of digits in numbers in terms of bills. Suppose you receive a paycheck for $564. Cashing the check at a bank, you receive a stack of bills—hundreds, tens, and ones. Sorted out, the stack looks something like this.

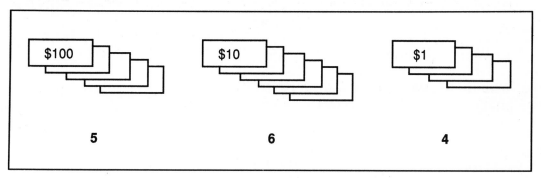

There are five $100 bills, six $10 bills, and four $1 bills. That's a good way to see that in the number 564 there are 5 hundreds, 6 tens, and 4 ones.

Zeros

Although the digit 0 has no numerical value, it holds a place in a number. For example, suppose your paycheck amounted to $504 instead of $564. Here is how the number 504 would look on a place-value chart.

Place Names	hundreds	tens	ones
Places	5	0	4
Place Values	100	10	1

The digit 4 is in the ones place. It has the value of 4 ones, or 4. The 0 in the tens place has the value of no tens, or 0. In other words, the 0 has no numerical value. However, it holds the tens place so that the digit 5 is in the hundreds place. The digit 5 has the value of 5 hundreds, or 500.

Without the zero as a placeholder, your $504 paycheck would be worth only $54!

EXERCISE 1a

Part A. Answer each question.

1. Which of the following are two-digit numbers?
   314   22   56   670   4

2. Which of the following are three-digit numbers?

523   142   34   987   5   900

3. Which digit is in the ones place in each number?
   (a) 12   (b) 3   (c) 402   (d) 89   (e) 98   (f) 6

4. Which digit is in the tens place in each number?
   (a) 12   (b) 67   (c) 312   (d) 831   (e) 509   (f) 33

5. Which digit is in the hundreds place in each number?
   (a) 505   (b) 678   (c) 943   (d) 142   (e) 200   (f) 607

6. What is the value of each underlined digit?
   (a) <u>5</u>7   (b) 84<u>6</u>   (c) <u>3</u>29   (d) 81<u>1</u>   (e) <u>5</u>   (f) 3<u>7</u>6

Part B. Complete each of the following statements.

1. 42 has ____ tens and ____ ones.

2. 31 has ____ tens and ____ one.

3. 93 has ____ tens and ____ ones.

4. 80 has ____ tens and ____ ones.

5. 55 has ____ tens and ____ ones.

6. 67 has ____ tens and ____ ones.

7. 689 has ____ hundreds, ____ tens, and ____ ones.

8. 401 has ____ hundreds, ____ tens, and ____ one.

9. 513 has ____ hundreds, ____ tens, and ____ ones.

10. 400 has ____ hundreds, ____ tens, and ____ ones.

11. 999 has ____ hundreds, ____ tens, and ____ ones.

12. 35 has ____ hundreds, ____ tens, and ____ ones.

Part C. Complete the following statements about the number 218.

1. The 8 is in the _____ place, so it has a value of _____ .

2. The 1 is in the _____ place, so it has a value of _____ .

3. The 2 is in the _____ place, so it has a value of _____ .

Part D. Complete the following statements about the number 706.

1. The 6 is in the _____ place, so it has a value of _____ .

2. The 0 is in the _____ place, and it has a value of _____ .

3. The 7 is in the _____ place, so it has a value of _____ .

Check your answers on page 206.

Thousands to Billions Places

You have learned the first three places of whole numbers. Here they are again with seven more whole number places—from the thousands to the billions.

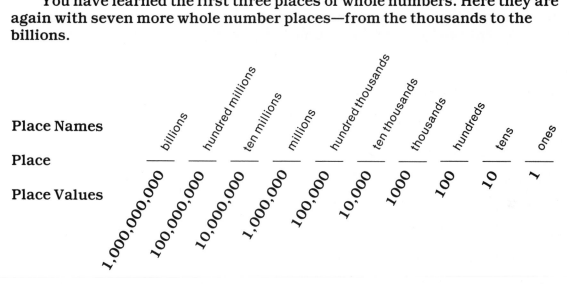

As you continue moving from right to left, the value of each place gets 10 times bigger.

Commas

Commas (,) separate the digits in whole numbers into groups of three digits. Commas have no value in themselves, but they make numbers easier to read. For example, according to a recent census, the number of people living in Chicago, Illinois, was 2,992,472. It is easier to read this large number with commas than without them: 2992472.

Commas are usually used in numbers with five digits or more, such as 62,109 and 1,430,112.

Commas can be used with four-digit numbers, but they are not necessary. For example, 3,200 and 3200 are both correct.

EXERCISE 1b

Part A. Answer each question.

1. Which digit is in the thousands place in each number?

   (a) 46,143 ___ (b) 5728 ___ (c) 56,340 ___

   (d) 7000 ___ (e) 19,037 ___ (f) 3406 ___

2. Which digit is in the ten-thousands place in each number?

    (a) 42,765 ___   (b) 89,001 ___   (c) 21,277 ___

    (d) 90,000 ___   (e) 76,142 ___   (f) 98,642 ___

3. What is the value of each underlined digit?

    (a) 4$\underline{5}$67 _____   (b) 1$\underline{1}$,070 _____

    (c) $\underline{1}$43,723 _____   (d) $\underline{8}$7,111 _____

    (e) $\underline{3}$0,123,654 _____   (f) $\underline{7}$,913,303,145 _____

Part B. In the place-value chart, write each of the following numbers. As an example, the first one is done for you.

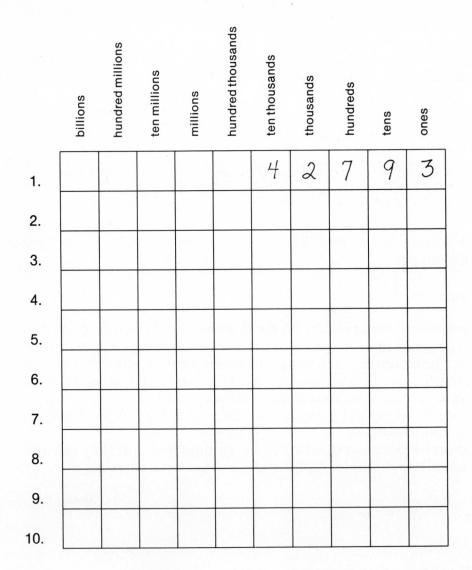

	billions	hundred millions	ten millions	millions	hundred thousands	ten thousands	thousands	hundreds	tens	ones
1.						4	2	7	9	3
2.										
3.										
4.										
5.										
6.										
7.										
8.										
9.										
10.										

1. 42,793   the number of books published in the United States in a recent year

2. 96,200   in dollars, the average price of a home in Atlanta, Georgia, in a recent year

3. 6185   the number of people who died in fires in the United States in a recent year

4. 31,827   the number of workers in the Upholsterers International Union of North America

5. 1459   the length, in miles, of the Arkansas River

6. 169,963,000   the number of eligible voters in the United States

7. 854,046   the number of books in the Wichita, Kansas, public library

8. 4141   the area, in square miles, of Hawaii

9. 2,009,683,000   in dollars, the annual budget of Rhode Island in a recent year

10. 93,000,000   the number of miles that the earth is from the sun

Part C. Complete each of the following statements.

1. 1536 has ____ thousand, ____ hundreds, ____ tens, and ____ ones.

2. 32,709 has ____ ten thousands, ____ thousands, ____ hundreds, ____ tens, and ____ ones.

3. 347,296 has ____ hundred thousands, ____ ten thousands, ____ thousands, ____ hundreds, ____ tens, and ____ ones.

4. 9,863,074 has ____ millions, ____ hundred thousands, ____ ten thousands, ____ thousands, ____ hundreds, ____ tens, and ____ ones.

5. 20,968,147 has ____ ten millions, ____ millions, ____ hundred thousands, ____ ten thousands, ____ thousands, ____ hundred, ____ tens, and ____ ones.

6. 3,298,645,009 has ____ billions, ____ hundred millions, ____ ten millions, ____ millions, ____ hundred thousands, ____ ten thousands, ____ thousands, ____ hundreds, ____ tens, and ____ ones.

Part D. Complete the following statements about the number 6804.

1. The 4 is in the _____ place, so it has a value of _____ .

2. The 0 is in the _____ place, and it has a value of _____ .

3. The 8 is in the _____ place, so it has a value of _____ .

4. The 6 is in the _____ place, so it has a value of _____ .

Part E. Complete the following statements about the number 8,902,365,741.

1. The 1 is in the _____ place, so it has a value of _____ .

2. The 4 is in the _____ place, so it has a value of _____ .

3. The 7 is in the _____ place, so it has a value of _____ .

4. The 5 is in the _____ place, so it has a value of _____ .

5. The 6 is in the _____ place, so it has a value of _____ .

6. The 3 is in the _____ place, so it has a value of _____ .

7. The 2 is in the _____ place, so it has a value of _____ .

8. The 0 is in the _____ place, and it has a value of _____ .

9. The 9 is in the _____ place, so it has a value of _____ .

10. The 8 is in the _____ place, so it has a value of _____ .

WORD PROBLEM

In one day in West Palm Beach, Florida, 19,032 copies of the local newspaper were sold. Identify the value of each digit in that number by writing it on a place-value chart.

Check your answers on page 206.

Lesson 2

Reading and Writing Whole Numbers

Whole numbers can be written in words.

Number (in digits)	Number (in words)
62	sixty-two
606	six hundred six
4,141	four thousand, one hundred forty-one
63,780	sixty-three thousand, seven hundred eighty
6,000,609	six million, six hundred nine
2,009,683,000	two billion, nine million, six hundred eighty-three thousand

Guidelines

When writing whole numbers, follow these guidelines.

- Numbers without zeros between 21 and 99 take a hyphen(-):

    twenty-one   sixty-four   ninety-nine

- A comma goes after the words *billion, million,* and *thousand* when they come in the middle of a number:

    four million, five hundred forty-five thousand, three hundred two

    BUT

    two billion
    one hundred five thousand

- Zeros hold places. They are not written in words:

20	twenty
200	two hundred
201	two hundred one
2010	two thousand ten

EXERCISE 2

Part A. Choose the correct answer for each item.

1. Which of the following is six thousand?
    - (1) 600
    - (2) 6000
    - (3) 60,000
    - (4) 600,000

2. Which of the following is two thousand, nineteen?
    - (1) 219
    - (2) 2019
    - (3) 2190
    - (4) 20,019

3. Which of the following is six hundred sixty-two thousand?
    - (1) 60,062
    - (2) 60,620
    - (3) 662,000
    - (4) 600,062,000

4. Which of the following is 4255?
    - (1) forty-two thousand, fifty-five
    - (2) forty thousand, two hundred fifty-five
    - (3) four thousand, fifty-five
    - (4) four thousand, two hundred fifty-five

5. Which of the following is 91,601?
    - (1) ninety-one thousand, six hundred one
    - (2) nine hundred thousand, six hundred one
    - (3) nine million, six hundred one
    - (4) nine million, six hundred one thousand

6. Which of the following is 4,077,000?
   (1) four hundred seventy-seven thousand
   (2) four million, seventy-seven thousand
   (3) four hundred thousand, seven thousand, seven
   (4) four billion, seventy-seven million

Part B. Write each number in digits. Be sure to put commas where they are needed. Use a place-value chart if it helps you see where to put zeros.

1. three million, four hundred eighteen thousand, six hundred twenty-three

   _____

2. six hundred forty-eight thousand _____

3. five million, two hundred thousand _____

4. three thousand, ten _____

5. ten thousand, four hundred forty-nine _____

6. thirty-three thousand, four hundred ninety-two _____

7. sixteen thousand, eight hundred fifty-six _____

8. forty-one thousand, nine hundred thirty-seven _____

9. eighty-five thousand, twenty _____

10. twenty-two thousand, two hundred twenty-one _____

Part C. Write each number in words. Be sure to put commas where they are needed.

1. 5240 _____

2. 8044 _____

3. 11,001 _____

4. 26,732 _____

5. 295,631 _____

6. 324,768 _____

7. 6,661,631 _____

8. 221,320,000 _____

9. 40,007,000,213 _____

10. 3,402,333,998 _____

Check your answers on page 207.

Ordering Whole Numbers

Lesson 3

To order whole numbers means to arrange them from largest to smallest or from smallest to largest.

Example: Put 231, 213, 312, 31, and 2213 in order from smallest to largest.

Step 1	Step 2	Step 3
231	231—③	31, 213, 231, 312, 2213
213	213—②	
312	312—④	
31	31—①	
2213	2213—⑤	

STEP 1: Write the numbers in a list with their digits in line according to place value.

STEP 2: By comparing the numbers, find the smallest and write a 1 next to it. Then find the next smallest, and write a 2 next to it. Continue until you have found the order for all of the numbers.

STEP 3: Rewrite the numbers with the smallest first, the next smallest second, and so on.

EXERCISE 3

Part A. Write the numbers in each set in order, from smallest to largest.

1. 238   4004   433   147   8   56

    ———— ,   ———— ,   ———— ,   ———— ,   ———— ,   ————

2. 82   915   60   555   6921   9000

____ , ____ , ____ , ____ , ____ , ____

3. 357   537   7002   2   2004   90

____ , ____ , ____ , ____ , ____ , ____

4. 4066   466   201   102   2001   500

____ , ____ , ____ , ____ , ____ , ____

Part B. Write the numbers in each set in order, from largest to smallest.

1. 71   52   93   342   314   324   43

____ , ____ , ____ , ____ , ____ , ____ , ____

2. 9   128   861   201   734   304   7034

____ , ____ , ____ , ____ , ____ , ____ , ____

3. 808   99   66   666   984   3894   894

____ , ____ , ____ , ____ , ____ , ____ , ____

4. 46   76   406   146   4006   820   4060

____ , ____ , ____ , ____ , ____ , ____ , ____

WORD PROBLEM

Arnie wants to make a list of the largest fish ever caught. His list will include a Spanish mackerel that weighed 12 pounds, a black sea bass that weighed 9 pounds, a white shark that weighed 2664 pounds, and a giant sea bass that weighed 563 pounds. If Arnie lists the fish from smallest to largest, in what order will they appear?

Check your answers on page 207.

Rounding Whole Numbers

People do not always use exact numbers. For example, a newspaper reporter might estimate that 30,000 people attended a concert, when there were really 29,768 people there. The reporter rounded the number 29,768 to the nearest ten thousand. The number of people at the concert was closer to 30,000 than to 29,000.

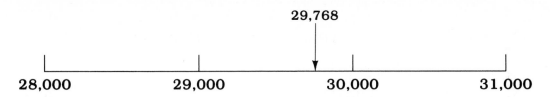

Many times an estimate is more useful than an exact amount. A contractor may estimate the cost of new aluminum siding for your house, for example. That is, the contractor would calculate the cost of materials and labor, using rounded numbers.

Follow these steps to round a whole number.

- Underline the digit in the place you want to round to.
- Is the digit to the right of the one you underlined less than 5? If it is, leave the underlined digit as it is. Is the digit to the right of the one you underlined 5 or more than 5? If it is, add 1 to the underlined digit.
- Change each digit to the right of the underlined digit to zero.

Example 1: Round $4322 to the nearest thousand dollars.

Step 1	Step 2	Step 3
		0 0 0
$4 3 2 5	$4 3 2 5	$4 3 2 5

STEP 1: Underline the digit in the thousands place—the 4.

STEP 2: Look at the digit to the right of the underlined 4. Since 3 is less than 5, leave the underlined digit as a 4.

STEP 3: Change the 3, the 2, and the 5, which are to the right of the underlined 4, to 0's. Rounded to the nearest thousand dollars, $4325 is $4000.

Example 2: Round $2173 to the nearest hundred dollars.

Step 1	Step 2	Step 3
	2	2 0 0
$2 1 7 3	$2 1 7 3	$2 1 7 3

STEP 1: Underline the digit in the hundreds place—the 1.

STEP 2: Look at the digit to the right of the underlined 1. Since 7 is more than 5, add 1 to the underlined 1, making it a 2.

STEP 3: Change the 7 and the 3, which are to the right of the underlined 1, to 0's. Rounded to the nearest hundred dollars, $2173 is $2200. $2200 is an estimate of the amount $2173.

Part A. Choose the correct answer to each of the following.

Rounded to the nearest ten,

1. 52 is	2. 74 is	3. 88 is	4. 254 is	5. 555 is
(1) 50	(1) 70	(1) 80	(1) 250	(1) 550
(2) 60	(2) 80	(2) 90	(2) 260	(2) 560

Rounded to the nearest hundred,

6. 549 is	7. $954 is
(1) 500	(1) $900
(2) 600	(2) $1000

Rounded to the nearest thousand,

8. 2907 is	9. 53,097 is
(1) 2000	(1) 53,000
(2) 3000	(2) 54,000

Rounded to the nearest hundred thousand,

10. 482,580 is

(1) 400,000

(2) 500,000

Part B. Round each number as directed.

1. Round 663 to the nearest hundred. _____

2. Round $549 to the nearest hundred dollars. _____

3. Round 2776 to the nearest thousand. _____

4. Round 7004 to the nearest thousand. _____

5. Round 91,999 to the nearest thousand. _____

6. Round 81,765 to the nearest ten. _____

7. Round 482,580 to the nearest hundred thousand. _____

8. Round 654,387 to the nearest hundred. _____

9. Round 8,458,903 to the nearest million. _____

10. Round 5643 to the nearest hundred. _____

11. Round 19,412 to the nearest ten thousand. _____

12. Round 44,999 to the nearest thousand. _____

Part C. Make estimates based on the following information about *Time* Magazine's circulation.

A recent report said that *Time* has a weekly circulation of 4,720,159.

1. Estimate *Time*'s circulation to the nearest million. _____

2. Estimate *Time*'s circulation to the nearest hundred thousand. _____

3. Estimate *Time*'s circulation to the nearest ten thousand. _____

WORD PROBLEM

Planning a sea voyage, Peter learned that the number of nautical, or ocean, miles between New Orleans, Louisiana, and Montreal, Quebec, is 3069. He estimated the distance to the nearest thousand and entered it in his notebook. What number did Peter write in the notebook?

Check your answers on page 208.

Chapter

2 ADDITION

The two lessons in this chapter cover the addition of whole numbers. The first lesson covers all the addition skills that do not involve carrying. The second covers carrying.

Lesson 5

Adding Whole Numbers

Adding is quicker than counting. For example, there are 3 tires in one stack and 4 tires in another stack. To find the total number of tires in both stacks, you can count them:

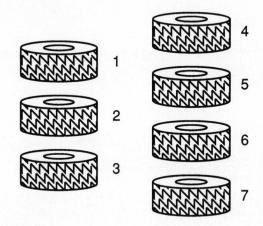

or you can add the numbers of tires in both stacks:

3 + 4 = 7

The sign for addition is a **plus sign** (+). This sign tells you to add two or more numbers. The answer to an addition problem is called the **sum**, or **total**. In the addition problem on page 22, the sum is 7.

The Basic Addition Facts

Knowing the basic addition facts will help you with any addition problem. These facts are listed in the Table of Basic Addition Facts on page 24. Learning these facts by heart will save you a lot of time. Take time to memorize one or two rows of the table each day until you know all of them by heart.

As you look over the table, notice these two things:

- The order in which numbers are added does not affect the sum. For example, $6 + 2 = 8$. If you reverse the order of the numbers, the answer is the same: $2 + 6 = 8$.
- When zero (0) is added to a number, the answer is the number itself. For example, $4 + 0 = 4$.

Adding One-Digit Numbers

Addition problems are often written across the page from left to right, like this:

$$3 + 4 = 7$$

However, it is usually easier to add digits if they are rewritten in a column like this:

$$\begin{array}{r} 3 \\ + 4 \\ \hline 7 \end{array}$$

Example 1:   Add $5 + 4$.

Step 1	Step 2
$\begin{array}{r} 5 \\ + 4 \\ \hline \end{array}$	$\begin{array}{r} 5 \\ + 4 \\ \hline 9 \end{array}$

STEP 1: Set up the problem in a column.

STEP 2: Add the numbers and write the sum. The answer is 9.

TABLE OF BASIC ADDITION FACTS

0 + 0 ─ 0	0 + 1 ─ 1	0 + 2 ─ 2	0 + 3 ─ 3	0 + 4 ─ 4	0 + 5 ─ 5	0 + 6 ─ 6	0 + 7 ─ 7	0 + 8 ─ 8	0 + 9 ─ 9
1 + 0 ─ 1	1 + 1 ─ 2	1 + 2 ─ 3	1 + 3 ─ 4	1 + 4 ─ 5	1 + 5 ─ 6	1 + 6 ─ 7	1 + 7 ─ 8	1 + 8 ─ 9	1 + 9 ─ 10
2 + 0 ─ 2	2 + 1 ─ 3	2 + 2 ─ 4	2 + 3 ─ 5	2 + 4 ─ 6	2 + 5 ─ 7	2 + 6 ─ 8	2 + 7 ─ 9	2 + 8 ─ 10	2 + 9 ─ 11
3 + 0 ─ 3	3 + 1 ─ 4	3 + 2 ─ 5	3 + 3 ─ 6	3 + 4 ─ 7	3 + 5 ─ 8	3 + 6 ─ 9	3 + 7 ─ 10	3 + 8 ─ 11	3 + 9 ─ 12
4 + 0 ─ 4	4 + 1 ─ 5	4 + 2 ─ 6	4 + 3 ─ 7	4 + 4 ─ 8	4 + 5 ─ 9	4 + 6 ─ 10	4 + 7 ─ 11	4 + 8 ─ 12	4 + 9 ─ 13
5 + 0 ─ 5	5 + 1 ─ 6	5 + 2 ─ 7	5 + 3 ─ 8	5 + 4 ─ 9	5 + 5 ─ 10	5 + 6 ─ 11	5 + 7 ─ 12	5 + 8 ─ 13	5 + 9 ─ 14
6 + 0 ─ 6	6 + 1 ─ 7	6 + 2 ─ 8	6 + 3 ─ 9	6 + 4 ─ 10	6 + 5 ─ 11	6 + 6 ─ 12	6 + 7 ─ 13	6 + 8 ─ 14	6 + 9 ─ 15
7 + 0 ─ 7	7 + 1 ─ 8	7 + 2 ─ 9	7 + 3 ─ 10	7 + 4 ─ 11	7 + 5 ─ 12	7 + 6 ─ 13	7 + 7 ─ 14	7 + 8 ─ 15	7 + 9 ─ 16
8 + 0 ─ 8	8 + 1 ─ 9	8 + 2 ─ 10	8 + 3 ─ 11	8 + 4 ─ 12	8 + 5 ─ 13	8 + 6 ─ 14	8 + 7 ─ 15	8 + 8 ─ 16	8 + 9 ─ 17
9 + 0 ─ 9	9 + 1 ─ 10	9 + 2 ─ 11	9 + 3 ─ 12	9 + 4 ─ 13	9 + 5 ─ 14	9 + 6 ─ 15	9 + 7 ─ 16	9 + 8 ─ 17	9 + 9 ─ 18

Set up the problems in columns. Find each sum.

1. $3 + 1 =$ 2. $4 + 5 =$ 3. $7 + 1 =$ 4. $4 + 2 =$

5. $6 + 2 =$ 6. $5 + 3 =$ 7. $6 + 1 =$ 8. $4 + 3 =$

9. $7 + 2 =$ 10. $3 + 3 =$ 11. $3 + 4 =$ 12. $2 + 4 =$

13. $2 + 1 =$ 14. $1 + 8 =$ 15. $2 + 7 =$ 16. $4 + 4 =$

17. $7 + 2 =$ 18. $1 + 1 =$ 19. $2 + 3 =$ 20. $5 + 4 =$

WORD PROBLEM

In word problems, look for key words that tell you how to solve the problem. For example, the words *in all* usually mean you must add.

Margie scored 2 goals in her first hockey game. Later that week she scored 3 goals in another game. How many goals **in all** did Margie score that week?

Check your answers on page 208.

Adding Two-Digit Numbers

When you add numbers with tens and ones places, you must line up digits in columns correctly. Put the ones under the ones, and the tens under the tens. Then add the numbers a column at a time, starting with the ones column.

Example 2:   Add $63 + 12$.

Step 1	**Step 2**	**Step 3**
tens ones	tens ones	tens ones
6 3	6 3	6 3
+ 1 2	+ 1 2	+ 1 2
	5	7 5

STEP 1: Set up the problem. Be sure that the digits are in the correct columns.

STEP 2: Add the numbers in the ones column: $3 + 2 = 5$.

STEP 3: Add the numbers in the tens column: $6 + 1 = 7$. The sum of 63 and 12 is 75.

Set up the problems in columns. Find each sum.

1. 52 + 33 =   2. 78 + 21 =   3. 41 + 38 =   4. 65 + 22 =

5. 43 + 34 =   6. 28 + 61 =   7. 13 + 54 =   8. 37 + 12 =

9. 11 + 23 =   10. 45 + 44 =   11. 17 + 21 =   12. 32 + 46 =

13. 52 + 34 =   14. 78 + 21 =   15. 61 + 36 =   16. 34 + 42 =

17. 53 + 35 =   18. 22 + 62 =   19. 31 + 41 =   20. 71 + 17 =

WORD PROBLEM

Remember that the key words *in all* tell you to add to solve the problem.

Fred worked 38 hours last week. This week he worked 11 hours. How many hours did he work **in all**?

Check your answers on page 208.

Adding One-Digit Numbers to Two-Digit Numbers

Sometimes, numbers with tens places are added to numbers with only ones places. When a place is empty, there is nothing to add.

Example 3:   Add $54 + $3.

Step 1	Step 2	Step 3
tens ones	tens ones	tens ones
$5 4	$5 4	$5 4
+ 3	+ 3	+ 3
	7	$5 7

STEP 1: Set up the problem. Be sure that the digits are in the correct columns.

STEP 2: Add the ones column: 4 + 3 = 7.

STEP 3: Add the tens column. There is no number under the 5. You can think of that place as zero: 5 + 0 = 5. The sum of $54 and $3 is $57.

In Example 3, the dollar sign ($) lets you know that you are adding two money amounts, *fifty-four dollars* and *three dollars*. When writing money amounts in an addition column, write the dollar sign ($) beside the first number and in the sum. You do not have to write a dollar sign beside other numbers in the column.

EXERCISE 5c

Set up the problems in columns. Find each sum.

1. 24 + 3 =   2. $35 + $4 =   3. 24 + 1 =   4. 55 + 4 =

5. $41 + $7 =   6. 63 + 6 =   7. 52 + 7 =   8. $98 + $1 =

9. 83 + 5 =   10. 92 + 3 =   11. $56 + $2 =   12. 33 + 6 =

13. $12 + $5 =   14. 22 + 7 =   15. 18 + 1 =   16. 31 + 5 =

17. 77 + 2 =   18. 91 + 4 =   19. $16 + $3 =   20. 62 + 6 =

WORD PROBLEM

In this word problem, the key word *altogether* tells you to add to solve the problem.

Juan has 46 CDs in his record collection. His sister Maria has 32. How many CDs do they have **altogether**?

Check your answers on page 208.

Adding Three-Digit Numbers

Like two-digit numbers, three-digit numbers are added a column at a time. First, add the ones column. Next, add the tens column. Finally, add the hundreds column.

Example 4: Find the sum of 756 and 213.

Step 1	Step 2	Step 3	Step 4
hundreds tens ones	hundreds tens ones	hundreds tens ones	hundreds tens ones
7 5 6	7 5 6	7 5 6	7 5 6
+ 2 1 3	+ 2 1 3	+ 2 1 3	+ 2 1 3
	9	6 9	9 6 9

STEP 1: Set up the problem. Be sure that the digits are in the correct columns.

STEP 2: Add the ones column: 6 + 3 = 9.

STEP 3: Add the tens column: 5 + 1 = 6.

STEP 4: Add the hundreds column: 7 + 2 = 9. The sum of 756 and 213 is 969.

EXERCISE 5d

Set up the problems in columns. Find each sum.

1. 836 + 142 =   2. 312 + 73 =   3. 531 + 427 =

4. 206 + 691 =   5. 426 + 351 =   6. $171 + $528 =

7. 230 + 8 =   8. 718 + 141 =   9. $602 + $101 =

10. 235 + 234 =   11. 123 + 634 =   12. 901 + 28 =

13. 522 + 436 =   14. 295 + 704 =   15. 830 + 53 =

16. 427 + 362 =   17. 551 + 40 =   18. 411 + 141 =

19. 600 + 180 =   20. 712 + 166 =

WORD PROBLEM

Look for the key words that tell you to add. Then solve the problem.

In an effort to stop overtime parking, the police began ticketing cars. The first month they ticketed 342 cars. The second month, they ticketed only 116 cars. How many cars in all did the police ticket?

Check your answers on page 208.

Adding Long Columns

To add a column with more than two numbers, find the sum of the first two numbers in the column. Then, add that sum to the next number in the column. Continue until you have added all the numbers in the column.

Example 5:   Find the sum of $3 + 2 + 1 + 2$.

Step 1	**Step 2**	**Step 3**	**Step 4**
3	3 ⎫	3 ⎫	3 ⎫
2	2 ⎭ $3 + 2 = 5$	2 ⎭ 5 ⎫	2 ⎭ 6 ⎫
1	1	1 ⎭ $5 + 1 = 6$	1 ⎭ $6 + 2 = 8$
+ 2	+ 2	+ 2	+ 2
			8

STEP 1: Set up the problem in a column.

STEP 2: Add the first two numbers: $3 + 2$. The sum is 5.

STEP 3: Add this sum to the next number: $5 + 1$. The new sum is 6.

STEP 4: Finally, add this new sum to the last number in the column: $6 + 2$. The answer is 8.

EXERCISE 5e

Set up the problems in columns. Find each sum.

1. $2 + 5 + 1 =$ 2. $1 + 4 + 2 + 2 =$ 3. $3 + 4 + 0 + 0 + 1 =$

4. $2 + 1 + 3 + 1 =$ 5. $5 + 2 + 1 + 1 =$ 6. $0 + 3 + 2 + 2 =$

7. $5 + 0 + 0 + 1 + 3 =$ 8. $3 + 4 + 2 =$ 9. $2 + 2 + 3 + 1 =$

10. $4 + 0 + 1 + 2 =$ 11. $3 + 3 + 3 =$ 12. $5 + 2 + 0 + 2 =$

WORD PROBLEM

In this word problem the key word *total* tells you to add to solve the problem.

Gerry played 4 baseball games in one week. In the first game he got 1 hit, and in the second game he got 3. He got 2 hits in the third game, but got no hits in the last game. What was the **total** number of hits Gerry got that week?

Check your answers on page 208.

Adding Long Columns of Two-Digit Numbers

To add three or more two-digit numbers, add the ones column first. Then add the tens column.

Example 6: Find the sum of $13 + 23 + 51$.

Step 1	Step 2	Step 3

$$
\begin{array}{r}
\text{tens ones} \\
1\ 3 \\
2\ 3 \\
+\ 5\ 1 \\
\end{array}
$$

Step 2:
$$
\begin{array}{r}
\text{tens ones} \\
1\ 3 \\
2\ 3 \\
+\ 5\ 1 \\
\hline
7
\end{array}
$$
$3 + 3 = 6$
$6 + 1 = 7$

Step 3:
$3 + 5 = 8$, $1 + 2 = 3$
$$
\begin{array}{r}
\text{tens ones} \\
1\ 3 \\
2\ 3 \\
+\ 5\ 1 \\
\hline
8\ 7
\end{array}
$$

STEP 1: Set up the problem. Be sure the digits are in their correct columns.

STEP 2: Add the first two numbers in the ones column: $3 + 3 = 6$. Add the sum to the third number in the ones column: $6 + 1 = 7$. Write the column sum, 7.

STEP 3: Add the first two numbers in the tens column: $1 + 2 = 3$. Add the sum to the third number in the tens column: $3 + 5 = 8$. Write the column sum, 8. The sum of $13 + 23 + 51$ is 87.

EXERCISE 5f

Set up the problems in columns. Find each sum.

1. $24 + 23 + 11 =$
2. $\$35 + \$33 + \$10 =$
3. $14 + 71 + 13 =$

4. $22 + 21 + 20 =$
5. $41 + 36 + 11 =$
6. $12 + 14 + 13 =$

7. $32 + 21 + 10 =$
8. $40 + 33 + 21 =$
9. $18 + 20 + 31 =$

10. $11 + 13 + 14 =$
11. $31 + 22 + 13 =$
12. $16 + 71 + 12 =$

13. $33 + 33 + 33 =$
14. $16 + 10 + 42 =$
15. $14 + 41 + 34 =$

16. $33 + 22 + 11 =$
17. $11 + 12 + 13 =$
18. $31 + 13 + 14 =$

19. $66 + 11 + 20 =$
20. $72 + 17 + 10 =$

WORD PROBLEM

Look for the key word that tells you to add. Then solve the problem.

Mr. Fiedler had his busiest day selling hot dogs last Tuesday. He sold 11 in the morning, 35 during the noon break, and another 43 before his day ended. What was the total number of hot dogs Mr. Fiedler sold that Tuesday?

Check your answers on page 208.

Adding Several Long Columns

No matter how large the numbers are, or how many columns there are to add, follow the same rules. Always add the ones column first. Then, add the tens column. Continue adding until you have added each column.

Example 7: Find the sum of 1335 + 5020 + 2623.

Step 1	Step 2	Step 3	Step 4	Step 5
thousands hundreds tens ones	thousands hundreds tens ones	thousands hundreds tens ones	thousands hundreds tens ones	thousands hundreds tens ones
1 3 3 5	1 3 3 5	1 3 3 5	1 3 3 5	1 3 3 5
5 0 2 0	5 0 2 0	5 0 2 0	5 0 2 0	5 0 2 0
+ 2 6 2 3	+ 2 6 2 3	+ 2 6 2 3	+ 2 6 2 3	+ 2 6 2 3
	8	7 8	9 7 8	8 9 7 8

STEP 1: Set up the problem. Be sure the digits are in their correct columns.

STEP 2: Add the numbers in the ones column: $5 + 0 = 5$; $5 + 3 = 8$.

STEP 3: Add the numbers in the tens column: $3 + 2 = 5$; $5 + 2 = 7$.

STEP 4: Add the numbers in the hundreds column: $3 + 0 = 3$; $3 + 6 = 9$.

STEP 5: Add the numbers in the thousands column: $1 + 5 = 6$; $6 + 2 = 8$. The answer is 8978.

EXERCISE 5g

Set up the problems in columns. Find each sum.

1. 4321 + 1234 + 3014 =

2. $6001 + $121 + $3042 =

3. 1352 + 4013 + 1223 =

4. 5060 + 1614 + 2314 =

5. 2173 + 4023 + 3603 =

6. 1234 + 4321 + 4 =

7. 7000 + 81 + 408 =

8. 5213 + 1302 + 461 =

9. 1032 + 2301 + 4156 =

10. 5111 + 222 + 3000 =

11. 2302 + 1431 + 5002 =

12. 9321 + 33 + 402 =

13. 1836 + 2151 + 3011 =

14. 4287 + 110 + 1 =

15. $3613 + $2024 + $1012 =

16. 5651 + 1022 + 2013 =

Check your answers on page 209.

Lesson
6

Carrying in Addition

When you add a column of numbers, the total may contain more than one place. When that happens, you need to carry.

Carrying to the Tens Column

In an addition problem, when you add the ones column, the total may have two digits. You need to carry the digit in the tens place of the total to the tens column of the addition problem. Then you add it with the other numbers in the tens column.

Example 1: Find the sum of 29 + 8.

Step 1	**Step 2**	**Step 3**
tens ones	tens ones	tens ones
2 9	2 9	2 9
+ 8	+ 8	+ 8
	7	3 7

STEP 1: Set up the problem.

STEP 2: Add the numbers in the ones column: $9 + 8 = 17$. Write the 7 in the ones place in the answer. Carry the 1 to the top of the tens column of the problem.

STEP 3: Add the numbers in the tens column, including the number you carried: $1 + 2 = 3$. Write the 3 in the tens place in the answer. The sum is 37.

Find the sums.

1. $16 + 8 =$	2. $29 + 3 =$	3. $\$38 + \$4 =$	4. $17 + 6 =$
5. $86 + 5 =$	6. $14 + 9 =$	7. $83 + 8 =$	8. $\$67 + \$7 =$
9. $24 + 6 =$	10. $9 + 2 =$	11. $32 + 9 =$	12. $\$3 + \$8 =$
13. $44 + 7 =$	14. $25 + 6 =$	15. $53 + 9 =$	16. $38 + 9 =$
17. $49 + 3 =$	18. $64 + 8 =$	19. $76 + 4 =$	20. $\$59 + \$9 =$

WORD PROBLEM

Find the key word that tells you to add. Then solve the problem.

During January, 17 inches of snow fell in Boulder, Colorado. In February, 8 more inches of snow fell on the town. What was the total number of inches of snow that fell in January and February?

Check your answers on page 209.

Carrying to Other Columns

Whenever the total of a column contains two digits, you need to carry. The following example shows carrying to the thousands column.

Example 2: Find the sum of $\$2735 + \$552$.

Step 1	**Step 2**	**Step 3**	**Step 4**	**Step 5**
thousands hundreds tens ones	thousands hundreds tens ones	thousands hundreds tens ones	thousands hundreds tens ones	thousands hundreds tens ones
			1	1
$\$2735$	$\$2735$	$\$2735$	$\$2735$	$\$2735$
$+552$	$+552$	$+552$	$+552$	$+552$
	7	87	287	$\$3287$

STEP 1: Set up the problem.

STEP 2: Add the numbers in the ones column: $5 + 2 = 7$. Write the 7 in the ones place in the answer.

STEP 3: Add the numbers in the tens column: 3 + 5 = 8. Write the 8 in the tens place in the answer.

STEP 4: Add the numbers in the hundreds column: 7 + 5 = 12. Write the 2 in the hundreds place in the answer. Carry the 1 to the top of the thousands column of the problem.

STEP 5: Add the numbers in the thousands column, including the number you carried: 1 + 2 = 3. Write the 3 in the thousands place in the answer. The sum is $3287.

EXERCISE 6b

Find the sums.

1. $71 + $45 =
2. 48 + 81 =
3. 294 + 94 =

4. 136 + 92 =
5. 785 + 74 =
6. 931 + 428 =

7. 622 + 706 =
8. $800 + $710 =
9. 263 + 63 =

10. 870 + 90 =
11. 1723 + 655 =
12. $962 + $704 =

13. 3312 + 734 =
14. 4540 + 603 =
15. 2858 + 1631 =

16. $1812 + $1712 =
17. 2709 + 3780 =
18. 4525 + 3552 =

19. 8722 + 322 =
20. 6816 + 2861 =

WORD PROBLEM

Combined is another key word that tells you to add. Read this word problem carefully. Then set up the problem and solve it.

Aldo has $715 in his account. His brother Ben has $624 in his account. How much money is in their accounts **combined**?

Check your answers on page 209.

Carrying More than Once

Sometimes you have to carry more than once in an addition problem.

Example 3: Find the sum of 28,975 + 3648.

Step 1	Step 2	Step 3	Step 4	Step 5

	ten thousands	thousands	hundreds	tens	ones

Step 1

$$
\begin{array}{r}
\;\;\;\;1\;\;\; \\
2\,8\,9\,7\,5 \\
+\;\;3\,6\,4\,8 \\
\hline
3
\end{array}
$$

Step 2

$$
\begin{array}{r}
\;\;\;1\;1\;\; \\
2\,8\,9\,7\,5 \\
+\;\;3\,6\,4\,8 \\
\hline
2\,3
\end{array}
$$

Step 3

$$
\begin{array}{r}
\;1\;1\;1\;\; \\
2\,8\,9\,7\,5 \\
+\;\;3\,6\,4\,8 \\
\hline
6\,2\,3
\end{array}
$$

Step 4

$$
\begin{array}{r}
1\;1\;1\;1\;\; \\
2\,8\,9\,7\,5 \\
+\;\;3\,6\,4\,8 \\
\hline
2\,6\,2\,3
\end{array}
$$

Step 5

$$
\begin{array}{r}
1\;1\;1\;1\;\; \\
2\,8\,9\,7\,5 \\
+\;\;3\,6\,4\,8 \\
\hline
3\,2\,6\,2\,3
\end{array}
$$

STEP 1: Add the numbers in the ones column: $5 + 8 = 13$. Write the 3 in the ones place and carry the 1 to the top of the tens column.

STEP 2: Add the numbers in the tens column, including the number you carried: $1 + 7 + 4 = 12$. Write the 2 in the tens place and carry the 1 to the top of the hundreds column.

STEP 3: Add the numbers in the hundreds column, including the number you carried: $1 + 9 + 6 = 16$. Write the 6 in the hundreds place and carry the 1 to the top of the thousands column.

STEP 4: Add the numbers in the thousands column, including the number you carried: $1 + 8 + 3 = 12$. Write the 2 in the thousands place and carry the 1 to the top of the ten-thousands column.

STEP 5: Add the numbers in the ten-thousands column, including the number you carried: $1 + 2 = 3$. Write the 3 in the ten-thousands place. The sum is 32,623.

EXERCISE 6c

Find each sum.

1. $549 + 372 =$
2. $985 + 257 =$
3. $\$234 + \$989 =$
4. $567 + 765 =$
5. $2849 + 3579 =$
6. $\$5648 + \$923 =$
7. $4837 + 5396 =$
8. $5876 + 1057 =$
9. $\$9287 + \$987 + \$602 =$
10. $6241 + 4876 + 79 =$
11. $457 + 387 + 545 =$
12. $763 + 658 + 123 =$
13. $\$12,041 + \$4,621 + \$57,193 =$
14. $13,522 + 6,491 + 46,208 =$
15. $12,795 + 986 + 213 =$
16. $64,441 + 3,999 + 312 =$

Find the key words that tell you to add. Then solve the problem.

Last summer Manny earned $2987 as a lifeguard. This summer he earned $3892. He has been offered the same job for next year at $4923. If Manny takes the job, how much money will he have earned in all for three summers as a lifeguard?

Check your answers on page 209.

Chapter

3 SUBTRACTION

The two lessons in this chapter cover the subtraction of whole numbers. The first lesson covers all the subtraction skills that do not involve borrowing. The second lesson covers borrowing.

Lesson 7

Subtracting Whole Numbers

Subtraction is *taking away* one number from another. In other words, subtraction is the opposite of addition. The sign for subtraction is a **minus sign** (−). Whenever you see this sign, you know that you must subtract. The answer to a subtraction problem is called the **difference**.

The Basic Subtraction Facts

Knowing the basic subtraction facts will help you with any subtraction problem. These facts are listed in the Table of Basic Subtraction Facts on page 38. Learning these facts by heart will save you a lot of time. Take time to memorize one or two rows of the table each day until you know all of them by heart.

As you look over the table, notice these three things:

- In basic subtraction, the larger number goes on top.
- When zero (0) is subtracted from a number, the answer is the number, itself. For example, $8 - 0 = 8$.
- When a number is subtracted from itself, the answer is zero (0). For example, $7 - 7 = 0$.

Subtracting One-Digit Numbers

As in addition, subtraction problems are often written from left to right, like this:

$$5 - 4 = 1$$

However, it is usually easier to subtract if you set up the problem in a column with the first number on top, like this:

$$\begin{array}{r} 5 \\ -4 \\ \hline 1 \end{array}$$

TABLE OF BASIC SUBTRACTION FACTS

0 −0 — 0	1 −0 — 1	2 −0 — 2	3 −0 — 3	4 −0 — 4	5 −0 — 5	6 −0 — 6	7 −0 — 7	8 −0 — 8	9 −0 — 9
1 −1 — 0	2 −1 — 1	3 −1 — 2	4 −1 — 3	5 −1 — 4	6 −1 — 5	7 −1 — 6	8 −1 — 7	9 −1 — 8	10 − 1 — 9
2 −2 — 0	3 −2 — 1	4 −2 — 2	5 −2 — 3	6 −2 — 4	7 −2 — 5	8 −2 — 6	9 −2 — 7	10 − 2 — 8	11 − 2 — 9
3 −3 — 0	4 −3 — 1	5 −3 — 2	6 −3 — 3	7 −3 — 4	8 −3 — 5	9 −3 — 6	10 − 3 — 7	11 − 3 — 8	12 − 3 — 9
4 −4 — 0	5 −4 — 1	6 −4 — 2	7 −4 — 3	8 −4 — 4	9 −4 — 5	10 − 4 — 6	11 − 4 — 7	12 − 4 — 8	13 − 4 — 9
5 −5 — 0	6 −5 — 1	7 −5 — 2	8 −5 — 3	9 −5 — 4	10 − 5 — 5	11 − 5 — 6	12 − 5 — 7	13 − 5 — 8	14 − 5 — 9
6 −6 — 0	7 −6 — 1	8 −6 — 2	9 −6 — 3	10 − 6 — 4	11 − 6 — 5	12 − 6 — 6	13 − 6 — 7	14 − 6 — 8	15 − 6 — 9
7 −7 — 0	8 −7 — 1	9 −7 — 2	10 − 7 — 3	11 − 7 — 4	12 − 7 — 5	13 − 7 — 6	14 − 7 — 7	15 − 7 — 8	16 − 7 — 9
8 −8 — 0	9 −8 — 1	10 − 8 — 2	11 − 8 — 3	12 − 8 — 4	13 − 8 — 5	14 − 8 — 6	15 − 8 — 7	16 − 8 — 8	17 − 8 — 9
9 −9 — 0	10 − 9 — 1	11 − 9 — 2	12 − 9 — 3	13 − 9 — 4	14 − 9 — 5	15 − 9 — 6	16 − 9 — 7	17 − 9 — 8	18 − 9 — 9

Example 1: Subtract 9 − 4.

Step 1	Step 2
9	9
− 4	− 4
	———
	5

STEP 1: Set up the problem. Be sure to put the first number on top.

STEP 2: Subtract 4 from 9. The answer is 5.

It is always a good idea to check subtraction problems. To do this, add the answer to the bottom number in the problem. The sum should equal the top number in the problem.

The following shows how to check the answer to the problem in Example 1.

Subtracting	**Checking**
9 top number	4 bottom number
− 4 bottom number	+ 5 answer
5 answer	9 top number

EXERCISE 7a

Set up each problem in a column. Subtract. Check your answers.

1. 3 − 2 =	2. 7 − 5 =	3. 5 − 1 =	4. $9 − $3 =
5. 6 − 4 =	6. $6 − $5 =	7. 8 − 3 =	8. 5 − 2 =
9. 9 − 7 =	10. 8 − 2 =	11. 7 − 2 =	12. 8 − 4 =
13. $9 − $0 =	14. 7 − 6 =	15. 6 − 3 =	16. 6 − 5 =
17. 9 − 1 =	18. 5 − 4 =	19. 7 − 1 =	20. 3 − 0 =

WORD PROBLEM

When you see the words *how much less* in a word problem, it is a signal to subtract.

Annette spent $8 for lunch on Monday. On Tuesday she spent $6. **How much less** did she spend on Tuesday?

Check your answers on page 209.

Subtracting Two-Digit Numbers

To subtract numbers with tens and ones places, begin by subtracting the numbers in the ones column. Then subtract the numbers in the tens column.

Example 2: Subtract 54 − 12.

Step 1	Step 2	Step 3	Step 4
tens ones	tens ones	tens ones	tens ones
5 4	5 4	5 4	1 2
− 1 2	− 1 2	− 1 2	+ 4 2
	2	4 2	5 4

STEP 1: Set up the problem. Be sure that the digits are in their correct columns.

STEP 2: Subtract the numbers in the ones column: 4 − 2 = 2.

STEP 3: Subtract the numbers in the tens column: 5 − 1 = 4. The answer is 42.

STEP 4: Add to check your answer: 12 + 42 = 54. The answer is correct.

EXERCISE 7b

Set up the problems in columns. Subtract. Check your answers.

1. 62 − 11 =   2. $35 − $24 =   3. 68 − 42 =   4. 97 − 63 =

5. 85 − 51 =   6. 28 − 12 =   7. 94 − 43 =   8. $37 − $16 =

9. 48 − 23 =   10. 76 − 33 =   11. 31 − 20 =   12. 44 − 22 =

13. $55 − $35 =   14. 48 − 23 =   15. 11 − 10 =   16. 89 − 70 =

17. 66 − 5 =   18. 56 − 24 =   19. $59 − $6 =   20. 85 − 33 =

> **WORD PROBLEM**
>
> The word *left* in the following problem tells you to subtract.
>
> Jack took $28 to the store. He spent $14 on a CD. How much did Jack have **left** after he paid for the CD?

Check your answers on page 209.

Subtracting Large Numbers

To subtract large numbers, subtract each column separately, beginning with the ones column.

Example 3: Subtract 7447 − 2436.

Step 1	Step 2	Step 3	Step 4	Step 5

| | thousands hundreds tens ones | thousands hundreds tens ones | thousands hundreds tens ones | thousands hundreds tens ones | thousands hundreds tens ones |

```
  Step 1        Step 2        Step 3        Step 4        Step 5

  7 4 4 7       7 4 4 7       7 4 4 7       7 4 4 7       2 4 3 6
− 2 4 3 6     − 2 4 3 6     − 2 4 3 6     − 2 4 3 6     + 5 0 1 1
        1            1 1         0 1 1       5 0 1 1       7 4 4 7
```

STEP 1: Set up the problem and subtract the numbers in the ones column: 7 − 6 = 1.

STEP 2: Subtract the numbers in the tens column: 4 − 3 = 1.

STEP 3: Subtract the numbers in the hundreds column: 4 − 4 = 0.

STEP 4: Subtract the numbers in the thousands column: 7 − 2 = 5.

STEP 5: Add to check your answer: 2436 + 5011 = 7447. The answer is correct.

EXERCISE 7c

Set up the problems in columns. Subtract. Check your answers.

1. 5642 − 4601 =
2. $7834 − $6423 =
3. 9567 − 4151 =

4. 4312 − 1102 =
5. 5974 − 4974 =
6. 9898 − 7676 =

7. 6459 − 3258 =
8. 2357 − 1245 =
9. $8971 − $7860 =

10. 7598 − 5491 =
11. 2998 − 1756 =
12. 4865 − 3555 =

13. $5742 − $311 =
14. 7727 − 6214 =
15. 8881 − 1531 =

16. 6850 − 4200 =
17. 3496 − 1284 =
18. 7569 − 52 =

19. 6822 − 620 =
20. $9534 − $8522 =

Check your answers on page 209.

Lesson 8

Borrowing in Subtraction

When you subtract a column of numbers, a digit in the top number may be smaller than the digit below it in the bottom number. When that happens, you need to borrow from the next column to the left in the top number.

Borrowing from the Tens Column

When the top number in the ones column is smaller than the bottom number in that column, you need to borrow 1 from the tens column. Because the 1 you borrow has a value of 10, you are borrowing 10 ones.

Example 1: Subtract $35 - 17$.

Step 1	Step 2	Step 3	Step 4
tens ones	tens ones	tens ones	tens ones
	2 15	2 15	1
$\begin{array}{r} 3\,5 \\ -\,1\,7 \\ \hline \end{array}$	$\begin{array}{r} 3\,\cancel{5} \\ -\,1\,7 \\ \hline 8 \end{array}$	$\begin{array}{r} 3\,\cancel{5} \\ -\,1\,7 \\ \hline 1\,8 \end{array}$	$\begin{array}{r} 17 \\ +\,18 \\ \hline 35 \end{array}$

STEP 1: Set up the problem in columns. Be sure to put the first number on top.

STEP 2: In the ones column, the top number, 5, is smaller than the number below it, 7. Borrow 1 from the 3 in the tens column. This leaves 2 in the tens column and gives you 15 in the ones column. Subtract the numbers in the ones column: $15 - 7 = 8$.

STEP 3: Subtract the numbers in the tens column: $2 - 1 = 1$. The answer is 18.

STEP 4: Add to check your answer: $17 + 18 = 35$. The answer is correct.

Set up the problems in columns. Be sure to put the first number on top. Subtract. Check each answer.

1. $18 − $9 =

2. 15 − 8 =

3. 13 − 4 =

4. 17 − 9 =

5. 11 − 3 =

6. 12 − 6 =

7. $14 − $5 =

8. 16 − 7 =

9. 15 − 6 =

10. 13 − 6 =

11. 43 − 19 =

12. 66 − 38 =

13. 74 − 55 =

14. $52 − $25 =

15. 57 − 49 =

16. 35 − 18 =

17. $88 − $59 =

18. 62 − 46 =

19. 95 − 56 =

20. $84 − $28 =

WORD PROBLEM

Look for the key word that tells you to subtract. Solve the problem.

Felix had $21 in his pocket. He paid $9 for a new pen knife. How much money did he have left?

Check your answers on page 209.

Borrowing Across a Zero

Sometimes the column you need to borrow from has a zero in the top number. Since you can't borrow from a zero, you need to borrow from the first column to the left that doesn't have a zero.

Example 2: Subtract 802 − 268.

Step 1	Step 2	Step 3	Step 4
hundreds tens ones	hundreds tens ones	hundreds tens ones	hundreds tens ones
	9 12	9 12	1 1
7 10	7 10	7 10	
8 0 2	8 0 2	8 0 2	2 6 8
− 2 6 8	− 2 6 8	− 2 6 8	+ 5 3 4
		5 3 4	8 0 2

STEP 1: Set up the problem in columns. You can't subtract the ones column without borrowing first, and you can't borrow from the 0 in the tens column. Borrow 1 from the 8 in the hundreds column. This leaves 7 in the hundreds column and gives you 10 in the tens column.

STEP 2: Borrow 1 from the 10 in the tens column. This leaves 9 in the tens column and gives you 12 in the ones column.

STEP 3: Subtract the numbers in each column. In the ones column, $12 - 8 = 4$; in the tens column, $9 - 6 = 3$; and in the hundreds column, $7 - 2 = 5$. The answer is 534.

STEP 4: Add to check your answer: $268 + 534 = 802$. The answer is correct.

EXERCISE 8b

Set up the problems in columns. Be sure to put the first number on top. Subtract. Check each answer.

1. $907 - 385 =$ 2. $\$605 - \$164 =$ 3. $505 - 253 =$

4. $306 - 147 =$ 5. $407 - 134 =$ 6. $202 - 103 =$

7. $307 - 152 =$ 8. $603 - 491 =$ 9. $808 - 527 =$

10. $909 - 111 =$ 11. $3407 - 1111 =$ 12. $6502 - 2391 =$

13. $7407 - 2116 =$ 14. $7804 - 2531 =$ 15. $3303 - 2121 =$

16. $8074 - 2133 =$ 17. $6063 - 1451 =$ 18. $3037 - 1712 =$

19. $6060 - 1310 =$ 20. $4033 - 3911 =$

WORD PROBLEM

In the following word problem, the words *how many more* are a clue that you need to subtract to find the answer.

The Do-Rite Company employed 4963 people last year. This year, the company employs 7086 people. **How many more** people does Do-Rite employ this year?

Check your answers on page 209.

Borrowing Across More than One Zero

Sometimes you may have to borrow across more than one zero.

Example 3: Subtract 38,006 − 12,629.

Step 1	Step 2	Step 3	Step 4	Step 5

STEP 1: Set up the problem in the columns. You can't subtract in the ones column without borrowing first, and you can't borrow from the zeros in the tens and hundreds columns. Borrow 1 from the 8 in the thousands column. This leaves 7 in the thousands column and gives you 10 in the hundreds column.

STEP 2: Borrow 1 from the 10 in the hundreds column. This leaves 9 in the hundreds column and gives you 10 in the tens column.

STEP 3: Borrow 1 from the 10 in the tens column. This leaves 9 in the tens column and gives you 16 in the ones column.

STEP 4: Subtract the numbers in each column. In the ones column, $16 − 9 = 7$; in the tens column, $9 − 2 = 7$; in the hundreds column, $9 − 6 = 3$; in the thousands column, $7 − 2 = 5$; and in the ten-thousands column, $3 − 1 = 2$. The answer is 25,377.

STEP 5: Add to check your answer: $12,626 + 25,377 = 38,006$. The answer is correct.

EXERCISE 8c

Set up the problems in columns. Be sure to put the first number on top. Subtract. Check each answer.

1. $600 − 233 =$ 2. $900 − 305 =$ 3. $400 − 242 =$

4. $300 − 99 =$ 5. $3600 − 1536 =$ 6. $2007 − 856 =$

7. $3008 − 1339 =$ 8. $3000 − 1053 =$ 9. $18,000 − 5,543 =$

10. $10,002 - 7,845 =$ 11. $30,018 - 1,197 =$ 12. $12,008 - 7,019 =$

13. $36,000 - 14,131 =$ 14. $355,000 - 432 =$

15. $400,205 - 3,016 =$ 16. $507,040 - 257,580 =$

WORD PROBLEM

The key word in this problem, *fewer*, tells you to subtract to solve the problem.

Last week a restaurant served 2000 meals. This week it served 1785 meals. How many **fewer** meals did it serve this week?

Check your answers on page 210.

Borrowing More than Once

To solve some subtraction problems, you need to borrow more than once.

Example 4: Subtract $934 - 768$.

Step 1	Step 2	Step 3	Step 4
hundreds tens ones	hundreds tens ones	hundreds tens ones	hundreds tens ones
2 14	8 12 / 2 14	8 12 / 2 14	1 1
9 3 4	9 3 4	9 3 4	7 6 8
− 7 6 8	− 7 6 8	− 7 6 8	+ 1 6 6
6	6 6	1 6 6	9 3 4

STEP 1: Set up the problem in columns. Borrow 1 from the 3 in the tens column to subtract the numbers in the ones column. Subtract: $14 - 8 = 6$.

STEP 2: Borrow 1 from the 9 in the hundreds column to subtract the numbers in the tens column. Subtract: $12 - 6 = 6$.

STEP 3: Subtract the numbers in the hundreds column: $8 - 7 = 1$. The answer is 166.

STEP 4: Add to check your answer: $768 + 166 = 934$. The answer is correct.

Set up the problems in columns. Be sure to put the first number on top. Subtract. Check each answer.

1. 547 − 248 =   2. 443 − 291 =   3. 115 − 86 =

4. 573 − 275 =   5. 375 − 186 =   6. 2222 − 43 =

7. 7509 − 2787 =   8. 5627 − 2179 =   9. 4582 − 1591 =

10. 6767 − 878 =   11. 8586 − 7397 =   12. 5674 − 2581 =

13. 7849 − 957 =   14. 8634 − 718 =   15. 2644 − 1445 =

16. 35,278 − 11,349 = 17. 29,685 − 9527 =   18. 19,833 − 3744 =

19. 84,947 − 5,930 = 20. 32,460 − 22,151 =

WORD PROBLEM

The key word *decreased* tells you to subtract.

At the start of a year, a new car was priced at $15,246, including sales tax. Later that year the same car was offered at $13,484, tax included. How much had the price of the car **decreased**?

Check your answers on page 210.

MIXED PRACTICE 1
ADDITION AND SUBTRACTION OF WHOLE NUMBERS

These problems let you practice the whole-number addition and subtraction skills you have learned so far. Read each problem carefully and solve it.

1. $3 + 4 =$ 　　　　 2. $78 - 24 =$ 　　　　 3. $5 + 2 =$

4. $37 - 6 =$ 　　　　 5. $56 - 24 =$ 　　　　 6. $8 + 7 =$

7. $9 + 0 =$ 　　　　 8. $30 - 20 =$ 　　　　 9. $3 + 7 =$

10. $23 - 3 =$ 　　　　 11. $77 + 22 =$ 　　　　 12. $43 + 6 =$

13. $289 - 208 =$ 　　　　 14. $916 - 915 =$ 　　　　 15. $91 + 4 =$

16. $18 + 61 =$ 　　　　 17. $541 - 240 =$ 　　　　 18. $43 + 34 =$

19. $76 - 48 =$ 　　　　 20. $936 - 524 =$ 　　　　 21. $55 - 46 =$

22. $72 + 47 =$ 　　　　 23. $63 + 90 =$ 　　　　 24. $20 - 19 =$

25. $92 + 92 =$ 　　　　 26. $63 - 56 =$ 　　　　 27. $88 - 39 =$

28. $80 + 50 =$ 　　　　 29. $603 - 202 =$ 　　　　 30. $4040 - 1210 =$

31. $77 + 51 =$ 　　　　 32. $812 + 156 =$ 　　　　 33. $602 + 103 =$

34. $808 - 709 =$ 　　　　 35. $6406 - 2245 =$ 　　　　 36. $816 + 147 =$

37. $909 + 104 =$ 　　　　 38. $42,005 - 19 =$ 　　　　 39. $3003 - 1414 =$

40. $600 - 279 =$ 　　　　 41. $461 + 426 =$ 　　　　 42. $5000 + 123 + 1000 =$

43. $60,008 - 4,129 =$ 　　　　 44. $4331 + 42 + 1130 =$

45. $12,000 - 5,508 =$ 　　　　 46. $3010 + 81 + 1214 =$

47. $60,563 - 1,942 =$ 　　　　 48. $6660 + 3124 + 407 =$

49. $2,000,000 - 296,357 =$ 　　　　 50. $2163 + 4545 + 2741 =$

Check your answers on page 210.

4 MULTIPLICATION

The two lessons in this chapter cover the multiplication of whole numbers. The first lesson covers most of the multiplication skills. The second covers shortcuts in multiplication.

Lesson 9

Multiplying Whole Numbers

Multiplication is a quick way to add the same number many times. For example, the following figure shows 4 rows of books with 7 books in each row.

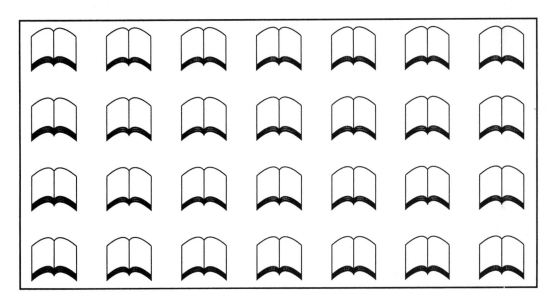

To find the total number of books, you can add the number of books in each row:

$$7 + 7 + 7 + 7 = 28$$

or you can multiply the number of rows by the number of books in each row:

$$4 \times 7 = 28$$

The sign for multiplication is a **multiplication sign**, or a **times sign** ($\times$). This sign tells you to multiply two numbers. The answer to a multiplication problem is called the **product**. In the multiplication problem above, the product is 28.

The Basic Multiplication Facts

Knowing the basic multiplication facts will help you with any multiplication problem. Learning the multiplication facts by heart will save you a lot of time. These facts are listed in The Multiplication Table that follows. Take time to memorize one or two rows of the table each day until you know the multiplication table by heart.

To find the product of two numbers in The Multiplication Table, read down from one number in the top row and across from the other number in the column on the far left. At the point where the lines meet, you'll find the product of the two numbers. For example, to multiply 4 and 7, find 4 in the top row of the table. Find 7 in the column on the far left. Read down from the 4 and across from the 7. The lines meet at 28, which is the product of the two numbers.

As you look over the table, notice these three things:

- The order in which numbers are multiplied does not change their product. For example, $6 \times 2 = 12$. If you reverse the order of the numbers, the answer is the same: $2 \times 6 = 12$.

- When a number is multiplied by zero (0), the product is zero (0). For example, $4 \times 0 = 0$.

- When a number is multiplied by 1, the product is the number itself. For example, $4 \times 1 = 4$.

THE MULTIPLICATION TABLE

	0	1	2	3	4	5	6	7	8	9	10	11	12
0	0	0	0	0	0	0	0	0	0	0	0	0	0
1	0	1	2	3	4	5	6	7	8	9	10	11	12
2	0	2	4	6	8	10	12	14	16	18	20	22	24
3	0	3	6	9	12	15	18	21	24	27	30	33	36
4	0	4	8	12	16	20	24	28	32	36	40	44	48
5	0	5	10	15	20	25	30	35	40	45	50	55	60
6	0	6	12	18	24	30	36	42	48	54	60	66	72
7	0	7	14	21	28	35	42	49	56	63	70	77	84
8	0	8	16	24	32	40	48	56	64	72	80	88	96
9	0	9	18	27	36	45	54	63	72	81	90	99	108
10	0	10	20	30	40	50	60	70	80	90	100	110	120
11	0	11	22	33	44	55	66	77	88	99	110	121	132
12	0	12	24	36	48	60	72	84	96	108	120	132	144

Multiply to find each product.

1. $4 \times 6 =$ 2. $6 \times 4 =$ 3. $10 \times 8 =$ 4. $8 \times 10 =$

5. $9 \times 8 =$ 6. $8 \times 9 =$ 7. $5 \times 7 =$ 8. $7 \times 5 =$

9. $11 \times 5 =$ 10. $5 \times 11 =$ 11. $2 \times 12 =$ 12. $12 \times 2 =$

13. $6 \times 3 =$ 14. $3 \times 6 =$ 15. $2 \times 7 =$ 16. $7 \times 2 =$

17. $4 \times 10 =$ 18. $10 \times 4 =$ 19. $12 \times 11 =$

20. $11 \times 12 =$ 21. $5 \times 0 =$ 22. $0 \times 3 =$

23. $1 \times 0 =$ 24. $8 \times 0 =$ 25. $2 \times 0 =$

26. $0 \times 4 =$ 27. $0 \times 2 =$ 28. $6 \times 0 =$

29. $0 \times 7 =$ 30. $9 \times 0 =$ 31. $5 \times 1 =$

32. $4 \times 1 =$ 33. $1 \times 3 =$ 34. $8 \times 1 =$

35. $1 \times 2 =$ 36. $9 \times 1 =$ 37. $7 \times 1 =$

38. $3 \times 1 =$ 39. $1 \times 8 =$ 40. $6 \times 1 =$

WORD PROBLEM

The key word *times* tells you to multiply.

Thelma Brown planted 8 cucumber plants in her garden. Next season she hopes to plant 3 **times** as many plants. How many cucumber plants will she need next season?

Check your answers on page 210.

Multiplying by One-Digit Numbers

To multiply a number by a one-digit number, first set up the problem in columns. Write the one-digit number on the bottom. Then, starting at the ones place, multiply each digit in the top number by the one-digit number.

Example 1: Multiply 13×2.

	Step 1	**Step 2**	**Step 3**
	tens ones	tens ones	tens ones
	1 3	1 3	1 3
	$\times$ 2	$\times$ 2	$\times$ 2
		6	2 6

STEP 1: Set up the problem in columns with the 2 on the bottom.

STEP 2: Multiply the 3 in the ones column of the top number by the bottom number: $3 \times 2 = 6$. Write the 6 in the ones place in the answer.

STEP 3: Multiply the 1 in the tens column of the top number by the bottom number: $1 \times 2 = 2$. Write the 2 in the tens place in the answer. The answer is 26.

EXERCISE 9b

Multiply to find each product.

1. $13 \times 3 =$	2. $21 \times 2 =$	3. $12 \times 2 =$	4. $30 \times 2 =$
5. $11 \times 7 =$	6. $23 \times 3 =$	7. $41 \times 2 =$	8. $22 \times 4 =$
9. $98 \times 1 =$	10. $14 \times 2 =$	11. $98 \times 0 =$	12. $16 \times 1 =$
13. $10 \times 6 =$	14. $34 \times 2 =$	15. $11 \times 6 =$	16. $222 \times 2 =$
17. $312 \times 3 =$	18. $111 \times 9 =$	19. $777 \times 1 =$	20. $111 \times 3 =$
21. $303 \times 3 =$	22. $114 \times 2 =$	23. $701 \times 0 =$	24. $333 \times 3 =$
25. $112 \times 4 =$			

WORD PROBLEM

As you know, the key words *in all* can be a clue that tells you to add. Because multiplication is a quick way to add the same number many times, the words *in all* can also tell you to multiply.

Stacia bought 12 used paperback books for $3 each. How much did she pay **in all** for the books?

Check your answers on page 210.

Carrying

When you multiply two numbers, you may need to carry.

Example 2: Multiply 43 × 6.

	Step 1	Step 2	Step 3
	tens ones	tens ones	tens ones
		1	1
	43	43	43
	× 6	× 6	× 6
		8	258

STEP 1: Set up the problem in columns with the 6 on the bottom.

STEP 2: Multiply the 3 in the ones column of the top number by the bottom number: $3 \times 6 = 18$. Write the 8 in the ones place in the answer and carry the 1 to the tens column of the problem.

STEP 3: Multiply the 4 in the tens column of the top number by the bottom number: $4 \times 6 = 24$. Add in the carried 1: $24 + 1 = 25$. Write the 5 in the tens place and the 2 in the hundreds place in the answer. The answer is 258.

EXERCISE 9c

Multiply to find each product.

1. $28 \times 4 =$
2. $19 \times 6 =$
3. $13 \times 4 =$

4. $28 \times 3 =$
5. $24 \times 4 =$
6. $16 \times 4 =$

7. $25 \times 3 =$
8. $39 \times 2 =$
9. $17 \times 5 =$

10. $73 \times 6 =$
11. $116 \times 7 =$
12. $140 \times 9 =$

13. $621 \times 5 =$
14. $691 \times 8 =$
15. $498 \times 3 =$

16. $208 \times 4 =$
17. $291 \times 6 =$
18. $571 \times 3 =$

19. $470 \times 8 =$
20. $740 \times 9 =$
21. $4124 \times 5 =$

22. $3032 \times 7 =$
23. $2412 \times 4 =$
24. $6230 \times 7 =$

25. $4132 \times 5 =$

If a problem asks you to find a total, you may be able to solve it by multiplication. Remember that multiplication is a quick way to add the same number many times.

Kay worked as a carpenter building homes for 7 months. She earned $2200 per month. What was her **total** pay for the 7 months?

Check your answers on page 210.

Multiplying by Two-Digit Numbers

To multiply by a two-digit number, multiply the top number by each digit in the bottom number. Multiply by the digit in the ones place of the bottom number first. Then multiply by the digit in the tens place.

Example 3: Multiply 42 × 21.

Step 1	Step 2	Step 3	Step 4
hundreds tens ones	hundreds tens ones	hundreds tens ones	hundreds tens ones
42	42	42	42
× 21	× 21	× 21	× 21
	42	42	42
		84	84
			882

STEP 1: Set up the problem in columns.

STEP 2: Multiply the top number by the 1 in the ones column of the bottom number: $2 \times 1 = 2$ and $4 \times 1 = 4$. Write the 2 in the ones column and the 4 in the tens column.

STEP 3: Multiply the top number by the 2 in the tens column of the bottom number: $2 \times 2 = 4$ and $4 \times 2 = 8$. Because you are multiplying by a number in the tens column, the 4 goes in the tens column of the answer and the 8 goes in the hundreds column.

STEP 4: Add the two answers to get the final product, 882.

Multiply to find each product.

1. $42 \times 12 =$
2. $30 \times 23 =$
3. $21 \times 20 =$

4. $40 \times 22 =$
5. $32 \times 11 =$
6. $22 \times 22 =$

7. $19 \times 10 =$
8. $21 \times 34 =$
9. $61 \times 10 =$

10. $12 \times 41 =$
11. $543 \times 56 =$
12. $568 \times 29 =$

13. $512 \times 14 =$
14. $444 \times 42 =$
15. $325 \times 43 =$

16. $408 \times 30 =$
17. $602 \times 32 =$
18. $314 \times 44 =$

19. $2815 \times 48 =$
20. $9520 \times 55 =$
21. $4397 \times 94 =$

22. $5403 \times 85 =$
23. $1122 \times 16 =$
24. $4321 \times 21 =$

25. $9876 \times 66 =$

WORD PROBLEM

The key word *at* followed by an amount of money can be a clue that you need to multiply. This problem gives you the price of one item and asks that you find the cost of more than one.

Ellen bought 11 pieces of costume jewelry **at** $28 each. What was the total amount of money she spent?

Check your answers on page 211.

Multiplying by Three-Digit Numbers

To multiply by a three-digit number, multiply the top number by each digit in the bottom number. Multiply by the digit in the ones place first. Then multiply by the digit in the tens place. Finally, multiply by the digit in the hundreds place.

Example 4: Multiply 352 × 274.

Step 1	Step 2	Step 3	Step 4

```
        Step 1              Step 2              Step 3              Step 4

                                                  1                   1
                              3 1                 3 1                 3 1
          2                   2                   2                   2
        3 5 2               3 5 2               3 5 2               3 5 2
      ×   2 7 4           ×   2 7 4           ×   2 7 4           ×   2 7 4
        1 4 0 8             1 4 0 8             1 4 0 8             1 4 0 8
                            2 4 6 4             2 4 6 4             2 4 6 4
                                                7 0 4               7 0 4
                                                                  9 6 4 4 8
```

STEP 1: Set up the problem in columns. Multiply the top number by the 4 in the bottom number: 4 × 2 = 8. Write the 8 in the ones column. 4 × 5 = 20: write the 0 in the tens column and carry the 2. 4 × 3 = 12; add the carried 2: 12 + 2 = 14. Write the 4 in the hundreds column and the 1 in the thousands column.

STEP 2: Multiply the top number by the 7 in the bottom number: 2 × 7 = 14. Write the 4 in the tens place and carry the 1. 5 × 7 = 35; add the carried 1: 35 + 1 = 36. Write the 6 in the hundreds place and carry the 3. 3 × 7 = 21; add the carried 3: 21 + 3 = 24. Write the 4 in the thousands place and the 2 in the ten-thousands place.

STEP 3: Multiply the top number by the 2 in the bottom number: 2 × 2 = 4: write the 4 in the hundreds place. 5 × 2 = 10: write the 0 in the thousands place and carry the 1. 3 × 2 = 6; add the carried 1: 6 + 1 = 7. Write the 7 in the ten-thousands place.

STEP 4: Add the three answers to get the final product, 96,448.

EXERCISE 9e

Multiply to find the products.

1. 221 × 314 =   2. 323 × 321 =   3. 142 × 121 =

4. 423 × 235 =   5. 497 × 118 =   6. 389 × 603 =

7. 227 × 650 =   8. 327 × 408 =   9. 2606 × 218 =

10. 2222 × 507 =   11. 1347 × 555 =   12. 8401 × 473 =

13. 5123 × 941 =   14. 4044 × 987 =   15. 2113 × 121 =

16. $1400 \times 212 =$ 17. $2727 \times 337 =$ 18. $6730 \times 243 =$

20. $4567 \times 411 =$

WORD PROBLEM

When a problem gives the price of one item and asks you to find the cost of many of those items, you can multiply to solve the problem.

Marcia Robles is providing the stoves for a new apartment complex. She is charging $468 each for the 231 stoves. What amount will she be paid for the stoves?

Check your answers on page 211.

Lesson 10

Shortcuts in Multiplication

Multiplying by Zero

There is a shortcut you can use when you multiply by a number with a zero.

Example 1: Multiply 216×30.

Step 1	Step 2	Step 3
thousands hundreds tens ones	thousands hundreds tens ones	thousands hundreds tens ones
$\begin{array}{r} 216 \\ \times\ \ 30 \\ \hline \end{array}$	$\begin{array}{r} 216 \\ \times\ \ 30 \\ \hline 0 \end{array}$	$\begin{array}{r} {\scriptstyle 1} \\ 216 \\ \times\ \ 30 \\ \hline 6480 \end{array}$

STEP 1: Set up the problem in columns.

STEP 2: Instead of multiplying 216 by 0 and writing three zeros, just write one zero under the zero in the bottom number.

STEP 3: Multiply 216 by the 3 in the bottom number. $6 \times 3 = 18$: write the 8 in the tens column and carry the 1. $1 \times 3 = 3$; add the carried 1: $3 + 1 = 4$. Write the 4 in the hundreds column. $2 \times 3 = 6$: write the 6 in the thousands column. The answer is 6480.

Multiply to find the products.

1. $21 \times 20 =$ 2. $33 \times 30 =$ 3. $64 \times 70 =$

4. $44 \times 40 =$ 5. $66 \times 80 =$ 6. $312 \times 30 =$

7. $233 \times 20 =$ 8. $112 \times 50 =$ 9. $897 \times 20 =$

10. $334 \times 70 =$ 11. $101 \times 40 =$ 12. $456 \times 20 =$

13. $632 \times 60 =$ 14. $999 \times 30 =$ 15. $711 \times 30 =$

16. $2121 \times 30 =$ 17. $4321 \times 30 =$ 18. $6789 \times 20 =$

19. $1261 \times 50 =$ 20. $3000 \times 20 =$

WORD PROBLEM

Remember that a key word for addition may be used in a problem you can solve by multiplying. Look for the key word in the following problem.

A major league team ordered 60 cases of baseballs. There are 144 baseballs in a case. How many baseballs did the team order altogether?

Check your answers on page 211.

You can use the same shortcut with any zero in the bottom number of a multiplication problem.

Example 2: Multiply 222×102.

Step 1	Step 2	Step 3	Step 4
$\begin{array}{r} 222 \\ \times\ 102 \\ \hline 444 \end{array}$	$\begin{array}{r} 222 \\ \times\ 102 \\ \hline 444 \\ 0 \end{array}$	$\begin{array}{r} 222 \\ \times\ 102 \\ \hline 444 \\ 2220 \end{array}$	$\begin{array}{r} 222 \\ \times\ 102 \\ \hline 444 \\ 2220 \\ \hline 22644 \end{array}$

STEP 1: Set up the problem in columns. Multiply the top number by the 2 in the ones place in the bottom number: $222 \times 2 = 444$.

STEP 2: Instead of multiplying 222 by 0 and writing 3 zeros, just write one zero under the zero in the bottom number.

STEP 3: Multiply 222 by the 1 in the hundreds place in the bottom number: $222 \times 1 = 222$. Write the answer next to the zero you wrote in Step 2.

STEP 4: Add the two answers to get the final product, 22,644.

EXERCISE 10b

Multiply to find the products.

1. $312 \times 203 =$ 2. $123 \times 101 =$ 3. $231 \times 402 =$

4. $554 \times 303 =$ 5. $897 \times 102 =$ 6. $418 \times 304 =$

7. $423 \times 201 =$ 8. $555 \times 405 =$ 9. $3299 \times 202 =$

10. $3434 \times 204 =$ 11. $666 \times 903 =$ 12. $456 \times 701 =$

13. $6221 \times 305 =$ 14. $2134 \times 503 =$ 15. $6432 \times 304 =$

16. $6321 \times 404 =$

WORD PROBLEM

Look for the key words that tell you to multiply to solve the problem.

For his office supply store, Mr. Whitman bought 102 cases of ballpoint pens at a cost of $133 per case. How much did he pay in all for the pens?

Check your answers on page 211.

Multiplying by 10, 100, and 1000

When you multiply a number by 10, 100, or 1000, you can use another shortcut. Follow these rules:

- To multiply a number by 10, write a zero (0) to the right of the number. For example, to multiply 22×10, just write a zero to the right of 22: 220.

- To multiply a number by 100, write two zeros (00) to the right of the number. For example, to multiply 22 × 100, just write two zeros to the right of 22: 2200.
- To multiply a number by 1000, write three zeros (000) to the right of the number. For example, to multiply 22 × 1000, just write three zeros to the right of 22: 22,000.

EXERCISE 10c

Multiply to find the products.

1. 10 × 36 =
2. 10 × 72 =
3. 34 × 10 =
4. 404 × 10 =
5. 67 × 10 =
6. 848 × 10 =
7. 765 × 10 =
8. 55 × 10 =
9. 100 × 3 =
10. 345 × 100 =
11. 100 × 702 =
12. 419 × 100 =
13. 1000 × 789 =
14. 4321 × 1000 =
15. 6279 × 1000 =
16. 1000 × 316 =
17. 10 × 423 =
18. 100 × 42 =
19. 1000 × 914 =
20. 100 × 6183 =

WORD PROBLEM

Look for the key word that tells you to multiply to solve the problem.

Alberta weighed the 75 boxes she was shipping to her customers. They each weighed exactly 100 pounds. What was the total weight of the 75 boxes?

Check your answers on page 211.

5 DIVISION

The two lessons in this chapter cover the division of whole numbers. The first lesson covers the basic division skills. The second covers division by larger numbers.

Lesson 11

Dividing Whole Numbers

Just as subtraction is the opposite of addition, division is the opposite of multiplication. The division signs are $\div$ and $\overline{)}$.

Division is a quick way to find out how many groups of the same size there are in a certain quantity. For example, if you want to know how many groups of 4 children there are in a class of 36, the easiest way to find out is to divide:

$$36 \div 4 = 9$$

The problem is read: 36 divided by 4 equals 9.

When you are working a division problem, it is best to set it up as follows:

$$4\overline{)36}^{\,9}$$

This problem is read: 4 divided into 36 equals 9.

Here are some terms to remember:
- The answer to a division problem is called the **quotient**.
- The number that divides into another number is called the **divisor**.
- The **dividend** is the number the divisor divides into.

$$\begin{array}{c} \textbf{divisor} \\ \downarrow \end{array}$$

$$\textbf{dividend} \rightarrow 36 \div 4 = 9 \leftarrow \textbf{quotient} \qquad \qquad \textbf{divisor} \rightarrow 4\overline{)36}^{\,9} \begin{array}{l} \leftarrow\textbf{quotient} \\ \leftarrow\textbf{dividend} \end{array}$$

The Basic Division Facts

Knowing the basic division facts will help you with any division problem. Learning the division facts by heart will save you a lot of time. These facts are listed in the Table of Basic Division Facts that follows. Take time each day to memorize one or two rows of the table until you know them by heart.

As you look over the table, notice these three things:

- When zero (0) is divided by a number, the quotient is zero (0). For example, $0 \div 7 = 0$.
- When a number is divided by 1, the quotient is the number itself. For example, $7 \div 1 = 7$.
- When a number is divided by itself, the quotient is 1. For example, $7 \div 7 = 1$.

TABLE OF BASIC DIVISION FACTS

$1\overline{)0}$ = 0	$1\overline{)1}$ = 1	$1\overline{)2}$ = 2	$1\overline{)3}$ = 3	$1\overline{)4}$ = 4	$1\overline{)5}$ = 5	$1\overline{)6}$ = 6	$1\overline{)7}$ = 7	$1\overline{)8}$ = 8	$1\overline{)9}$ = 9
$2\overline{)0}$ = 0	$2\overline{)2}$ = 1	$2\overline{)4}$ = 2	$2\overline{)6}$ = 3	$2\overline{)8}$ = 4	$2\overline{)10}$ = 5	$2\overline{)12}$ = 6	$2\overline{)14}$ = 7	$2\overline{)16}$ = 8	$2\overline{)18}$ = 9
$3\overline{)0}$ = 0	$3\overline{)3}$ = 1	$3\overline{)6}$ = 2	$3\overline{)9}$ = 3	$3\overline{)12}$ = 4	$3\overline{)15}$ = 5	$3\overline{)18}$ = 6	$3\overline{)21}$ = 7	$3\overline{)24}$ = 8	$3\overline{)27}$ = 9
$4\overline{)0}$ = 0	$4\overline{)4}$ = 1	$4\overline{)8}$ = 2	$4\overline{)12}$ = 3	$4\overline{)16}$ = 4	$4\overline{)20}$ = 5	$4\overline{)24}$ = 6	$4\overline{)28}$ = 7	$4\overline{)32}$ = 8	$4\overline{)36}$ = 9
$5\overline{)0}$ = 0	$5\overline{)5}$ = 1	$5\overline{)10}$ = 2	$5\overline{)15}$ = 3	$5\overline{)20}$ = 4	$5\overline{)25}$ = 5	$5\overline{)30}$ = 6	$5\overline{)35}$ = 7	$5\overline{)40}$ = 8	$5\overline{)45}$ = 9
$6\overline{)0}$ = 0	$6\overline{)6}$ = 1	$6\overline{)12}$ = 2	$6\overline{)18}$ = 3	$6\overline{)24}$ = 4	$6\overline{)30}$ = 5	$6\overline{)36}$ = 6	$6\overline{)42}$ = 7	$6\overline{)48}$ = 8	$6\overline{)54}$ = 9
$7\overline{)0}$ = 0	$7\overline{)7}$ = 1	$7\overline{)14}$ = 2	$7\overline{)21}$ = 3	$7\overline{)28}$ = 4	$7\overline{)35}$ = 5	$7\overline{)42}$ = 6	$7\overline{)49}$ = 7	$7\overline{)56}$ = 8	$7\overline{)63}$ = 9
$8\overline{)0}$ = 0	$8\overline{)8}$ = 1	$8\overline{)16}$ = 2	$8\overline{)24}$ = 3	$8\overline{)32}$ = 4	$8\overline{)40}$ = 5	$8\overline{)48}$ = 6	$8\overline{)56}$ = 7	$8\overline{)64}$ = 8	$8\overline{)72}$ = 9
$9\overline{)0}$ = 0	$9\overline{)9}$ = 1	$9\overline{)18}$ = 2	$9\overline{)27}$ = 3	$9\overline{)36}$ = 4	$9\overline{)45}$ = 5	$9\overline{)54}$ = 6	$9\overline{)63}$ = 7	$9\overline{)72}$ = 8	$9\overline{)81}$ = 9

Divide to find each quotient.

1. $4\overline{)28}$	2. $2\overline{)14}$	3. $8\overline{)16}$	4. $2\overline{)0}$	5. $8\overline{)48}$
6. $7\overline{)7}$	7. $3\overline{)27}$	8. $2\overline{)6}$	9. $9\overline{)36}$	10. $1\overline{)3}$
11. $7\overline{)42}$	12. $9\overline{)45}$	13. $4\overline{)20}$	14. $2\overline{)4}$	15. $3\overline{)21}$
16. $3\overline{)9}$	17. $5\overline{)10}$	18. $7\overline{)21}$	19. $9\overline{)54}$	20. $8\overline{)0}$
21. $4\overline{)32}$	22. $7\overline{)49}$	23. $2\overline{)18}$	24. $5\overline{)15}$	25. $8\overline{)72}$
26. $5\overline{)40}$	27. $6\overline{)12}$	28. $3\overline{)3}$	29. $5\overline{)20}$	30. $1\overline{)9}$
31. $5\overline{)35}$	32. $8\overline{)64}$	33. $3\overline{)24}$	34. $3\overline{)12}$	35. $9\overline{)63}$
36. $4\overline{)8}$	37. $1\overline{)8}$	38. $8\overline{)40}$	39. $6\overline{)36}$	40. $7\overline{)63}$
41. $3\overline{)15}$	42. $2\overline{)2}$	43. $5\overline{)0}$	44. $6\overline{)48}$	45. $4\overline{)16}$
46. $4\overline{)28}$	47. $9\overline{)27}$	48. $4\overline{)0}$	49. $6\overline{)30}$	50. $3\overline{)6}$
51. $9\overline{)72}$	52. $2\overline{)12}$	53. $2\overline{)10}$	54. $4\overline{)36}$	55. $2\overline{)16}$
56. $4\overline{)24}$	57. $9\overline{)81}$	58. $7\overline{)35}$	59. $2\overline{)8}$	60. $8\overline{)56}$
61. $9\overline{)0}$	62. $3\overline{)18}$	63. $7\overline{)14}$	64. $5\overline{)45}$	65. $1\overline{)5}$
66. $6\overline{)18}$	67. $8\overline{)32}$	68. $5\overline{)25}$	69. $6\overline{)54}$	70. $6\overline{)42}$
71. $5\overline{)5}$	72. $6\overline{)24}$	73. $8\overline{)8}$	74. $6\overline{)0}$	75. $1\overline{)9}$
76. $5\overline{)30}$	77. $7\overline{)56}$	78. $8\overline{)24}$	79. $1\overline{)7}$	80. $9\overline{)18}$

WORD PROBLEM

The key word *each* can be a signal that you need to divide to solve a problem.

Four people contributed equal amounts of money to buy a birthday cake for James. The cake cost $24. How much did **each** person contribute toward the price of the cake?

Check your answers on page 211.

Dividing by One-Digit Numbers

To divide by a one-digit number, first set up the problem. Then divide the divisor into the dividend one digit at a time, beginning with the place farthest to the left and working toward the right. Check the quotient by multiplying it times the divisor. If you have divided correctly, the answer will be the same as the dividend.

Example 1: Divide 693 ÷ 3.

Step 1	Step 2	Step 3	Step 4	Step 5
$3\overline{)693}$	$\begin{array}{r}2\\3\overline{)693}\end{array}$	$\begin{array}{r}23\\3\overline{)693}\end{array}$	$\begin{array}{r}231\\3\overline{)693}\end{array}$	$\begin{array}{r}231\\\times\quad 3\\\hline 693\end{array}$

STEP 1: Set up the problem.

STEP 2 Divide the 3 into the 6 in the hundreds place of the dividend: 6 ÷ 3 = 2. Write the 2 in the hundreds place in the answer.

STEP 3: Divide the 3 into the 9 in the tens place of the dividend: 9 ÷ 3 = 3. Write the 3 in the tens place of the answer.

STEP 4: Divide the 3 into the 3 in the ones place of the dividend: 3 ÷ 3 = 1. Write the 1 in the ones place of the answer. The quotient is 231.

STEP 5: Check your answer: multiply the quotient, 231, by the divisor, 3. The answer is the same as the dividend, 693.

EXERCISE 11b

Divide to find each quotient. Check your answers.

1. 48 ÷ 2 =   2. 84 ÷ 4 =   3. 55 ÷ 5 =   4. 64 ÷ 2 =

5. 36 ÷ 6 =   6. 93 ÷ 3 =   7. 44 ÷ 2 =   8. 66 ÷ 3 =

9. 84 ÷ 4 =   10. 96 ÷ 3 =   11. 633 ÷ 3 =   12. 488 ÷ 4 =

13. 468 ÷ 2 =   14. 242 ÷ 2 =   15. 848 ÷ 4 =   16. 642 ÷ 2 =

17. 999 ÷ 9 =   18. 700 ÷ 7 =   19. 888 ÷ 4 =   20. 636 ÷ 3 =

21. 4884 ÷ 4 =   22. 4466 ÷ 2 =   23. 3636 ÷ 3 =   24. 8642 ÷ 2 =

25. 9999 ÷ 9 =

Check your answers on page 211.

Dividing into Smaller Digits

Sometimes you can't divide into the first digit in a dividend because it is smaller than the divisor. Then you must divide into the first two digits of the dividend.

Example 2:   Divide $155 \div 5$.

Step 1	Step 2	Step 3	Step 4
hundreds tens ones	hundreds tens ones	hundreds tens ones	hundreds tens ones
$5\overline{)1\,5\,5}$	$\begin{array}{r} 3 \\ 5\overline{)1\,5\,5} \end{array}$	$\begin{array}{r} 3\,1 \\ 5\overline{)1\,5\,5} \end{array}$	$\begin{array}{r} 3\,1 \\ \times\ \ 5 \\ \hline 1\,5\,5 \end{array}$

STEP 1: Set up the problem.

STEP 2: You can't divide 5 into the 1 in the hundreds place of the divisor. Therefore, you need to divide into the first two places of the divisor: $15 \div 5 = 3$. Write the 3 in the tens place in the answer.

STEP 3: Divide 5 into the 5 in the ones place of the dividend: $5 \div 5 = 1$. Write the 1 in the ones place of the answer. The quotient is 31.

STEP 4: Check your answer. Multiply the quotient, 31, by the divisor, 5. The answer is the same as the dividend, 155.

EXERCISE 11c

Divide to find each quotient. Check your answers.

1. $255 \div 5 =$
2. $249 \div 3 =$
3. $124 \div 2 =$
4. $368 \div 4 =$

5. $546 \div 6 =$
6. $186 \div 3 =$
7. $248 \div 8 =$
8. $328 \div 4 =$

9. $279 \div 3 =$
10. $486 \div 6 =$
11. $217 \div 7 =$
12. $205 \div 5 =$

13. $819 \div 9 =$
14. $288 \div 4 =$
15. $166 \div 2 =$
16. $3288 \div 8 =$

17. $1648 \div 4 =$ 18. $3555 \div 5 =$ 19. $1569 \div 3 =$

20. $1446 \div 2 =$ 21. $2799 \div 9 =$ 22. $2466 \div 3 =$

23. $3248 \div 4 =$ 24. $5466 \div 6 =$ 25. $2488 \div 8 =$

WORD PROBLEM

In the following problem, look for the word that tells you to divide to solve the problem.

Charley packed 328 paperback books in 4 boxes. He put an equal number of books into each box. How many books did he put into each box?

Check your answers on page 212.

Zeros in Quotients

Sometimes you can't divide into the second or another digit in a dividend because it is smaller than the divisor. Then you must put a zero in the quotient above that digit to hold the place.

Example 3: Divide $812 \div 4$.

Step 1	Step 2	Step 3	Step 4
hundreds tens ones	hundreds tens ones	hundreds tens ones	hundreds tens ones
2	2 0	2 0 3	2 0 3
$4\overline{)8\ 1\ 2}$	$4\overline{)8\ 1\ 2}$	$4\overline{)8\ 1\ 2}$	$\times\quad 4$
			8 1 2

STEP 1: Set up the problem. Divide 4 into the 8 in the hundreds place of the dividend: $8 \div 4 = 2$. Write the 2 in the hundreds place in the quotient.

STEP 2: You can't divide 4 into the 1 in the tens place of the divisor. Write a zero (0) in the tens place in the quotient to hold the place.

STEP 3: Divide into two places of the divisor: $12 \div 4 = 3$. Write the 3 in the ones place in the quotient. The quotient is 203.

STEP 4: Check your answer: multiply the quotient, 203, by the divisor, 4. The answer is the same as the dividend, 812.

Divide to find each quotient. Check your answers.

1. $624 \div 3 =$	2. $820 \div 4 =$	3. $510 \div 5 =$	4. $812 \div 2 =$
5. $918 \div 9 =$	6. $618 \div 6 =$	7. $808 \div 4 =$	8. $612 \div 2 =$
9. $520 \div 5 =$	10. $707 \div 7 =$	11. $315 \div 3 =$	12. $828 \div 4 =$
13. $515 \div 5 =$	14. $2828 \div 7 =$	15. $1821 \div 3 =$	16. $6060 \div 6 =$
17. $6129 \div 3 =$	18. $8804 \div 4 =$	19. $5155 \div 5 =$	20. $4144 \div 2 =$
21. $3069 \div 3 =$	22. $8206 \div 2 =$	23. $4124 \div 4 =$	24. $8168 \div 8 =$
25. $5255 \div 5 =$			

WORD PROBLEM

Find the key word that tells you to divide to solve the following problem.

Jessica is an engineer. She received a fee of $6276 for work she did on a new bridge. Her fee was divided into three equal payments. What was the amount of each payment?

Check your answers on page 212.

Long Division

In the division problems you have worked so far, the divisors have divided into the dividends evenly. Sometimes, however, a divisor does not divide evenly into a dividend. When this happens, you need to use long division. In long division, you repeat these four steps until you have found the quotient:

- Estimate and/or divide.
- Multiply.
- Subtract and compare.
- Bring down the next number.

Estimating is usually an important part of long division. For example, you may need to divide 5 into 18. It won't divide evenly. You need to remember what numbers close to 18 can be divided evenly by 5. You remember that $20 \div 5 = 4$. You realize, however, that 5 will not divide into 18 four times. You remember that $15 \div 5 = 3$. Therefore, you can figure that 5 will divide into 18 three times with some left over.

Example 4: Divide 180 ÷ 5.

Step 1	Step 2	Step 3	Step 4	Step 5	Step 6	Step 7
hundreds tens ones	hundreds tens ones	hundreds tens ones	hundreds tens ones	hundreds tens ones	hundreds tens ones	hundreds tens ones

$$
\begin{array}{r} 3 \\ 5\overline{)180} \end{array}
\qquad
\begin{array}{r} 3 \\ 5\overline{)180} \\ \underline{15} \end{array}
\qquad
\begin{array}{r} 3 \\ 5\overline{)180} \\ \underline{15} \\ 3 \end{array}
\qquad
\begin{array}{r} 3 \\ 5\overline{)180} \\ \underline{15} \\ 30 \end{array}
\qquad
\begin{array}{r} 36 \\ 5\overline{)180} \\ \underline{15} \\ 30 \end{array}
\qquad
\begin{array}{r} 36 \\ 5\overline{)180} \\ \underline{15} \\ 30 \\ 30 \end{array}
\qquad
\begin{array}{r} 36 \\ 5\overline{)180} \\ \underline{15} \\ 30 \\ \underline{30} \\ 0 \end{array}
$$

STEP 1: Set up the problem. **Estimate** how many times 5 will divide into the first two places of the dividend: 5 will not divide into 18 four times, but it will divide into 18 three times. **Divide:** 18 ÷ 5 = 3 with some left over. Write the 3 in the tens place in the quotient.

STEP 2: **Multiply** the 3 in the quotient by the divisor: 3 × 5 = 15. Write the 15 under the 18 in the dividend.

STEP 3: **Subtract** the 15 you wrote from the 18: 18 − 15 = 3. **Compare** the 3 with the divisor. It should be smaller than the divisor. Since 3 is smaller than 5, go to Step 4.

STEP 4: **Bring down the next number** from the dividend, which is 0. Write it beside the 3.

STEP 5: **Divide** 5 into 30: 30 ÷ 5 = 6. (You do not need to estimate this time because 5 divides into 30 evenly.) Write the 6 in the ones place in the quotient.

STEP 6: **Multiply** the 6 in the quotient by the divisor: 6 × 5 = 30. Write that 30 under the 30 you divided into.

STEP 7: **Subtract:** 30 − 30 = 0. Because the answer is 0, there is nothing left to divide. The problem is finished. The quotient is 36. To check, multiply the quotient by the divisor: 36 × 5 = 180.

EXERCISE 11e

Use long division to solve these problems.

1. 315 ÷ 5 =   2. 441 ÷ 7 =   3. 414 ÷ 9 =   4. 170 ÷ 5 =

5. 324 ÷ 6 =   6. 117 ÷ 9 =   7. 148 ÷ 2 =   8. 243 ÷ 3 =

9. 497 ÷ 7 =   10. 544 ÷ 8 =   11. 100 ÷ 5 =   12. 380 ÷ 5 =

13. 476 ÷ 7 =   14. 136 ÷ 4 =   15. 264 ÷ 6 =   16. 640 ÷ 8 =

17. $465 \div 5 =$ 18. $231 \div 3 =$ 19. $405 \div 5 =$

20. $117 \div 3 =$ 21. $5944 \div 8 =$ 22. $1248 \div 2 =$

23. $4571 \div 7 =$ 24. $5544 \div 6 =$ 25. $1134 \div 9 =$

WORD PROBLEM

Find the key word that tells you to divide to solve the following problem.

In one week 258 new recruits reported for basic training with the Marine Corps. The recruits were divided into 6 training platoons. How many recruits were in each training platoon?

Check your answers on page 212.

Remainders

Division problems do not always come out evenly. When that happens, there is a **remainder**. Use the letter **r** to show the part of the quotient that is the remainder.

Example 5: Divide $37 \div 6$.

Step 1	Step 2	Step 3	Step 4	Step 5

$$
\begin{array}{ccccc}
\text{Step 1} & \text{Step 2} & \text{Step 3} & \text{Step 4} & \text{Step 5} \\
& & & & \\
5\overline{)37} & 5\overline{)37} & \begin{array}{r} 7 \\ 5\overline{)37} \\ 35 \\ \hline 2 \end{array} & \begin{array}{r} 7\,r\,2 \\ 5\overline{)37} \\ 35 \\ \hline 2 \end{array} & \begin{array}{r} 7 \\ \times\ 5 \\ \hline 35 \\ +\ 2 \\ \hline 37 \end{array}
\end{array}
$$

STEP 1: Set up the problem.

STEP 2: You can't divide 5 into 3, so you need to divide into both places of the dividend: $37 \div 5 = 7$ with some left over. Write the 7 in the ones place in the answer.

STEP 3: Multiply the 7 in the quotient times the divisor: $7 \times 5 = 35$. Write the 35 under the 37, the dividend.

STEP 4: Subtract: $37 - 35 = 2$. Compare the 2 with the divisor. Because it is smaller than the divisor, and there are no numbers to bring down, the problem is finished. The remainder is 2. Write it, following an **r** as part of the quotient. The answer is 7 r 2.

STEP 5: To check the answer: multiply the quotient by the divisor and add the remainder: $7 \times 5 = 35$; $35 + 2 = 37$. The answer is the same as the dividend.

Find each quotient. Check your answers.

1. $7 \div 3 =$ 2. $9 \div 4 =$ 3. $7 \div 2 =$ 4. $6 \div 4 =$

5. $11 \div 5 =$ 6. $50 \div 7 =$ 7. $75 \div 9 =$ 8. $24 \div 5 =$

9. $26 \div 4 =$ 10. $50 \div 8 =$ 11. $79 \div 3 =$ 12. $35 \div 8 =$

13. $59 \div 7 =$ 14. $19 \div 4 =$ 15. $89 \div 8 =$ 16. $49 \div 3 =$

17. $75 \div 6 =$ 18. $98 \div 8 =$ 19. $65 \div 4 =$ 20. $79 \div 5 =$

21. $723 \div 7 =$ 22. $555 \div 2 =$ 23. $973 \div 4 =$ 24. $628 \div 5 =$

25. $728 \div 6 =$

WORD PROBLEM

The key word *equal* can be a clue that you need to divide to solve a problem.

Ruben, a taxicab driver, drove 985 miles in 3 days. He drove an **equal** number of miles each day. How many miles did he drive each day?

Check your answers on page 212.

Lesson 12

Dividing by Two- and Three-Digit Numbers

Two-Digit Divisors

Dividing by a two-digit number is like dividing by a one-digit number. You have to estimate how many times the divisor (the two-digit number) will go into the dividend. If your first guess is wrong, just guess again.

Example 1: Divide 880 ÷ 20.

Step 1	Step 2	Step 3	Step 4	Step 5

Step 1:
```
      hundreds
      │ tens
      │ │ ones
20 ) 8 8 0
```

Step 2:
```
          4
20 ) 8 8 0
```

Step 3:
```
          4
20 ) 8 8 0
      8 0
      ───
        8
```

Step 4:
```
         4 4
20 ) 8 8 0
      8 0
      ───
      8 0
```

Step 5:
```
         4 4
20 ) 8 8 0
      8 0
      ───
      8 0
      8 0
      ───
         0
```

STEP 1: Set up the problem.

STEP 2: Divide 20 into the first two places of the dividend. Estimate the answer by thinking this way: 20 divided into 80 equals 4, so 20 divided into 88 will equal 4 with some left over. Write the 4 in the tens place in the quotient.

STEP 3: Multiply the 4 by the divisor: $4 \times 20 = 80$. Write 80 under the 88 in the dividend and subtract: $88 - 80 = 8$. Compare the 8 with the divisor. It is smaller than the divisor, so your estimate was right.

STEP 4: Bring down the 0 from the dividend, and write it beside the 8. Divide: $80 \div 20 = 4$. Write the 4 in the ones place in the quotient.

STEP 5: Multiply: $4 \times 20 = 80$. Write that 80 under the 80 you divided into and subtract. Because the answer is 0, there is nothing left to divide. The problem is finished. The quotient is 44. To check, multiply the quotient by the divisor: $44 \times 20 = 880$.

EXERCISE 12a

Divide to find each quotient. Some answers will have remainders. Check your answers.

1. $270 \div 54 =$ 2. $733 \div 81 =$ 3. $344 \div 43 =$ 4. $108 \div 36 =$

5. $988 \div 82 =$ 6. $60 \div 10 =$ 7. $75 \div 15 =$ 8. $33 \div 11 =$

9. $78 \div 14 =$ 10. $63 \div 25 =$ 11. $144 \div 36 =$ 12. $270 \div 54 =$

13. $160 \div 40 =$ 14. $344 \div 43 =$ 15. $120 \div 24 =$ 16. $441 \div 72 =$

17. $763 \div 81 =$ 18. $250 \div 61 =$ 19. $267 \div 44 =$ 20. $245 \div 35 =$

21. $289 \div 71 =$ 22. $204 \div 51 =$ 23. $190 \div 24 =$ 24. $164 \div 47 =$

25. $396 \div 52 =$

The key words *divide*, *evenly*, and *each* tell you to use division to solve this problem.

Mr. Alvarez, the president of the P.T.A., knew that 525 parents would be attending the meetings. If Mr. Alvarez **divides** the parents **evenly** among 25 meeting rooms, how many will be in **each** room?

Check your answers on page 212.

Dividing into Large Numbers

You can use the method you just learned to solve any division problem. Keep dividing until you find the answer. Then check by multiplying.

Example 2: Divide 8692 ÷ 41.

Step 1	Step 2	Step 3	Step 4	Step 5

STEP 1: Set up the problem.

STEP 2: Divide: 86 ÷ 41. You can estimate that the answer is 2 because 80 ÷ 40 = 2. Write a 2 in the hundreds place in the quotient. Multiply: 2 × 41 = 82. Write the 82 under the 86. Subtract and compare: 86 − 82 = 4, which is less than the divisor, 41, so the estimate was correct.

STEP 3: Bring down the 9, making 49. Divide: 49 ÷ 41 = 1. Write the 1 in the quotient. Multiply: 1 × 41 = 41. Write the 41 under the 49. Subtract: 49 − 41 = 8.

STEP 4: Bring down the 2, making 82. Divide: 82 ÷ 41 = 2. Write the 2 in the quotient. Multiply: 2 × 41 = 82. Subtract: 82 − 82 = 0. There is no remainder.

STEP 5: Multiply to check your answer.

Divide to find each quotient. Some answers will have remainders. Check your answers.

1. $8867 \div 42 =$

2. $3808 \div 34 =$

3. $5776 \div 25 =$

4. $6099 \div 19 =$

5. $6535 \div 15 =$

6. $8700 \div 15 =$

7. $7280 \div 70 =$

8. $4633 \div 32 =$

9. $7044 \div 24 =$

10. $1764 \div 28 =$

11. $4899 \div 23 =$

12. $6168 \div 34 =$

13. $8270 \div 73 =$

14. $2208 \div 46 =$

15. $8500 \div 40 =$

16. $20{,}237 \div 49 =$

17. $34{,}848 \div 66 =$

18. $60{,}066 \div 44 =$

19. $55{,}555 \div 25 =$

20. $88{,}016 \div 40 =$

21. $87{,}000 \div 75 =$

22. $13{,}026 \div 39 =$

23. $46{,}884 \div 24 =$

24. $22{,}092 \div 42 =$

25. $60{,}423 \div 33 =$

WORD PROBLEM

Look for the key words that tell you to divide to solve this problem.

On opening night, a movie was shown in 35 theaters across the country. Altogether, 4375 people attended the movie that night. If the people were divided evenly among the theaters, how many people saw the movie in each theater?

Check your answers on page 212.

Three-Digit Divisors

Dividing by three digits is similar to dividing by two digits. Use the same method; the difference is in the size of the numbers.

Example 3: Divide 8997 ÷ 214.

Step 1	Step 2	Step 3	Step 4

Step 1

thousands hundreds tens ones

$$214\overline{)8997}$$ with quotient 4

$$\begin{array}{r} 4 \\ 214\overline{)8997} \\ 856 \\ \hline 43 \end{array}$$

Step 2

$$\begin{array}{r} 42 \\ 214\overline{)8997} \\ 856 \\ \hline 437 \\ 428 \\ \hline 9 \end{array}$$

Step 3

$$\begin{array}{r} 42\,r\,9 \\ 214\overline{)8997} \\ 856 \\ \hline 437 \\ 428 \\ \hline 9 \end{array}$$

Step 4

$$\begin{array}{r} 214 \\ \times\quad 42 \\ \hline 428 \\ 856 \\ \hline 8988 \\ +\quad\ 9 \\ \hline 8997 \end{array}$$

STEP 1: Set up the problem and divide: 899 ÷ 214. You can estimate that the answer is 4 because 800 ÷ 200 = 4. Multiply: 4 × 214 = 856. Write the answer under the 899. Subtract and compare: 899 − 856 = 43, which is less than the divisor, 214, so the estimate was correct.

STEP 2: Bring down the 7, making 437. Divide: 437 ÷ 214 = 2. Multiply: 2 × 214 = 428. Write the answer under the 437. Subtract: 437 − 428 = 9.

STEP 3: 9 can't be divided by 214, and there are no more numbers to bring down. Write r 9 in the answer.

STEP 4: Multiply and then add to check the answer.

EXERCISE 12c

Divide and check your answers.

1. 974 ÷ 323 =	2. 564 ÷ 140 =	3. 735 ÷ 365 =
4. 658 ÷ 324 =	5. 384 ÷ 128 =	6. 1256 ÷ 289 =
7. 1900 ÷ 176 =	8. 6986 ÷ 225 =	9. 9988 ÷ 467 =
10. 3047 ÷ 145 =	11. 2120 ÷ 424 =	12. 1525 ÷ 502 =
13. 3047 ÷ 145 =	14. 6002 ÷ 342 =	15. 6939 ÷ 390 =
16. 3340 ÷ 250 =	17. 7244 ÷ 223 =	18. 8422 ÷ 222 =
19. 1010 ÷ 505 =	20. 9420 ÷ 456 =	21. 25,505 ÷ 225 =
22. 17,388 ÷ 322 =	23. 46,672 ÷ 418 =	24. 14,570 ÷ 170 =

WORD PROBLEM

Look for the key words that tell you to divide to solve the following problem.

Four-Wheel Motors gave each of its 120 employees a bonus last month. Each employee got the same amount of money. The company paid out $23,520. How much was each employee's bonus?

Check your answers on page 212.

MIXED PRACTICE 2
MULTIPLICATION AND DIVISION OF WHOLE NUMBERS

These problems let you practice the whole-number multiplication and division skills you have learned so far. Read each problem carefully and solve it.

1. $42 \times 4 =$ 2. $16 \div 2 =$ 3. $12 \times 4 =$

4. $5 \div 5 =$ 5. $28 \times 0 =$ 6. $72 \div 9 =$

7. $56 \div 8 =$ 8. $49 \div 7 =$ 9. $14 \times 3 =$

10. $20 \times 1 =$ 11. $366 \div 3 =$ 12. $5010 \div 5 =$

13. $673 \times 6 =$ 14. $204 \times 7 =$ 15. $1414 \div 7 =$

16. $4682 \times 3 =$ 17. $6666 \div 6 =$ 18. $848 \div 4 =$

19. $5348 \times 4 =$ 20. $1625 \times 8 =$ 21. $36 \times 12 =$

22. $1570 \div 5 =$ 23. $1866 \div 6 =$ 24. $2484 \div 4 =$

25. $40 \times 31 =$ 26. $503 \times 32 =$ 27. $216 \times 28 =$

28. $4806 \div 6 =$ 29. $3627 \div 9 =$ 30. $77 \times 12 =$

31. $622 \times 214 =$ 32. $811 \times 411 =$ 33. $35 \div 8 =$

34. $82 \div 9 =$ 35. $14 \div 3 =$ 36. $227 \div 20 =$

37. $725 \times 253 =$ 38. $434 \times 343 =$ 39. $645 \div 32 =$

40. $5050 \div 25 =$ 41. $650 \times 823 =$ 42. $2002 \times 154 =$

43. $5423 \times 232 =$ 44. $3264 \div 32 =$ 45. $4384 \div 42 =$

46. $6006 \times 300 =$ 47. $9127 \times 447 =$ 48. $22,626 \div 63 =$

49. $3465 \times 0 =$ 50. $48,109 \div 24 =$ 51. $624 \div 312 =$

52. $2926 \div 434 =$ 53. $1824 \times 243 =$ 54. $4000 \times 372 =$

55. $62,345 \times 304 =$ 56. $21,203 \times 107 =$ 57. $80,880 \div 800 =$

58. $44,678 \div 654 =$ 59. $7453 \times 234 =$ 60. $32,455 \div 205 =$

Check your answers on page 212.

WHOLE-NUMBER SKILLS REVIEW

Part A. Add, subtract, multiply, or divide.

1. $24 \times 9 =$
2. $75 - 37 =$
3. $66 + 28 =$

4. $437 - 68 =$
5. $476 \times 8 =$
6. $872 + 564 =$

7. $365 \times 26 =$
8. $24 \div 8 =$
9. $4795 - 2963 =$

10. $908 + 683 =$
11. $92 \div 4 =$
12. $357 + 63 =$

13. $450 \div 18 =$
14. $607 \times 54 =$
15. $5987 - 654 =$

16. $1735 \div 32 =$
17. $802 \times 21 =$
18. $777 - 385 =$

19. $682 + 68 =$
20. $500 - 34 =$
21. $311 - 24 =$

22. $963 + 874 =$
23. $241 \times 60 =$
24. $180 \div 6 =$

25. $3471 - 2980 =$
26. $624 \times 402 =$
27. $100 \div 20 =$

28. $477 + 88 =$
29. $322 \div 14 =$
30. $300 \times 89 =$

Part B. Solve each problem.

1. Denise must divide 384 bags of apples among 24 boxes. How many bags should she pack in each box?

2. Last week Esteban earned $478 at his regular job. He also earned $169 working part time for his uncle. How much did he earn in all?

3. Carmella works as an inspector in a clothing factory. On Monday she inspected 2154 blouses. She inspected 1746 blouses on Tuesday. How many more blouses did she inspect on Monday than on Tuesday?

4. Chan's Cantonese Cafe ordered 36 boxes of teacups. At $12 per box, how much did Chan's pay for the teacups altogether?

5. If 183 roses are divided evenly among 61 centerpieces, how many roses will be in each centerpiece?

6. If 12 full-time employees at a lumberyard each earn $15,800 per year, how much does the lumberyard spend each year for their salaries?

7. A part-time bus driver, Henry worked 3 days last week. On Friday he drove 227 miles. On Saturday he drove 341 miles. On Sunday he drove another 287 miles. How many miles did Henry drive in all?

8. Henry's gross earnings for driving the bus were $240. His employer withheld $56 from his paycheck for taxes. What was the amount of Henry's pay after taxes?

Check your answers on page 213.

Chapter

6 USING WHOLE NUMBERS

In the seven lessons in this chapter, you will learn to solve problems that involve measurements, perimeter and area, and averages. You will also learn to solve one-step and multistep word problems.

Lesson 13

Units of Measure

How tall are you? How far is it to work? How much milk should you put in the pancake batter? how much flour? How many days, hours, and minutes is it until your next day off? All these questions are about measuring something.

Every country has basic measures for ordinary things. Many countries use the metric system. In the United States, we use the English system, which was brought here many years ago. In this lesson you will work with the English and the metric systems.

Measuring Length

Listed below are measures for length in the English and metric systems.

ENGLISH	METRIC
12 inches (in.) = 1 foot (ft)	1000 millimeters (mm) = 1 meter (m)
3 feet = 1 yard (yd)	100 centimeters (cm) = 1 meter
36 inches = 1 yard	10 decimeters (dm) = 1 meter
5280 feet = 1 mile (mi)	10 meters = 1 dekameter (dam)
1760 yards = 1 mile	10 dekameter = 1 hectometer (hm)
	10 hectometers = 1 kilometer (km)
	1000 meters = 1 kilometer

The English and metric systems are different. The metric system is based on multiples of 10. To change from one unit of measure to another, you multiply or divide by 10. The English system uses several different multiples. For example, to change between feet and inches, you multiply or divide by 12.

To change from a larger unit of length measure to a smaller unit, multiply the number of larger units by the number of smaller units in one larger unit. (This rule works with every kind of measuring unit in the English and the metric systems.)

Example 1: Change 2 feet to inches.

Step 1	**Step 2**
1 ft = 12 in.	12 in. $\times$ 2 24 in.

STEP 1: Find the number of inches in one foot. There are 12 inches in 1 foot.

STEP 2: Multiply the number of inches in 1 foot, 12, by the number of feet you have, 2: 2 feet equals 24 inches.

Example 2: Change 2 meters to centimeters.

Step 1	**Step 2**
1 m = 100 cm	100 cm $\times$ 2 200 cm

STEP 1: Find the number of centimeters in 1 meter. There are 100 cm in 1 m.

STEP 2: Multiply the number of centimeters in one meter, 100, by the number of meters you have, 2: 2 meters equals 200 centimeters.

EXERCISE 13a

Multiply to change each unit of length to the one shown.

1. 9 ft = _____ in.
2. 4 m = _____ mm
3. + mi = _____ yd
4. 7 dam = _____ m
5. 108 yd = _____ in.
6. 1 dam = _____ m
7. 2 km = _____ m
8. 43 yd = _____ ft
9. 3 km = _____ hm
10. 4 yd = _____ in.

WORD PROBLEM

Answer the following question.

Peter wants to put a shelf under the window in his daughter's room. He wants the shelf to be as wide as the window, which is 48 inches. He has a board 4 feet long. Is it long enough to serve as the shelf?

Check your answers on page 213.

To change a smaller unit of a measure to a larger unit, divide the number of smaller units by the number of smaller units in one larger unit. (This rule works with any English or metric unit of measure.)

Example 3: Change 72 inches to yards.

Step 1

1 yd = 36 in.

Step 2

$$\begin{array}{r} 2 \text{ yd} \\ 36\overline{)72 \text{ in.}} \\ \underline{72} \\ 0 \end{array}$$

STEP 1: Find the number of inches in 1 yard. There are 36 inches in 1 yard.

STEP 2: Divide the number of inches you have, 72, by the number of inches in one yard, 36: 72 inches equals 2 yards.

Example 4: Change 200 centimeters to meters.

Step 1

1 m = 100 cm

Step 2

$$\begin{array}{r} 2 \text{ m} \\ 100\overline{)200 \text{ cm}} \\ \underline{200} \\ 0 \end{array}$$

STEP 1: Find the number of centimeters in 1 meter. There are 100 cm in 1 m.

STEP 2: Divide the number of centimeters you have, 200, by the number of centimeters in 1 meter, 100: 200 centimeters equals 2 meters.

EXERCISE 13b

Divide to change each unit of length to the one shown.

1. 36 in. = _____ yd
2. 40 dm = _____ m
3. 180 in. = _____ ft
4. 129 ft = _____ yd
5. 30 hm = _____ km
6. 30 ft = _____ yd
7. 4000 m = _____ km
8. 10,560 ft = _____ mi
9. 80 dam = _____ hm
10. 60 m = _____ dam

WORD PROBLEM

Solve the following problem.

Lee has a plumber's snake 108 inches long. What is the length of the snake in feet?

Check your answers on page 213.

Measuring Weight

English measures of weight are different from metric measures. This chart shows measures of weight in both systems.

ENGLISH	METRIC

ENGLISH

16 ounces (oz) = 1 pound (lb)
2000 pounds = 1 ton (T)

METRIC

1000 milligrams (mg) = 1 gram (g)
100 centigrams (cg) = 1 gram
10 decigrams (dg) = 1 gram
10 grams = 1 dekagram (dag)
10 dekagrams = 1 hectogram (hg)
10 hectograms = 1 kilogram (kg)
1000 grams = 1 kilogram

The metric system has more ways to describe weights than the English system does. The metric system is used in medicine because it is useful in describing very small amounts. For example, the label on a medicine container may show that the medicine is measured in milligrams.

To change one unit of weight to another, use the same rules you learned for length: Multiply to change from larger to smaller units, and divide to change from smaller to larger units.

EXERCISE 13c

Change each unit of weight to the one shown.

1. 2 T = _____ lb
2. 2 g = _____ mg
3. 96 oz = _____ lb
4. 12 lb = _____ oz
5. 20 hg = _____ kg
6. 10,000 lb = _____ T
7. 32 oz = _____ lb
8. 5 T = _____ lb
9. 3 g = _____ dg
10. 64 oz = _____ lb
11. 4 g = _____ cg
12. 16,000 lb = _____ T

WORD PROBLEM

Answer the following question.

According to his recipe for spaghetti sauce, Jason needs 1 pound of mushrooms. In a vegetable store, he finds mushrooms packaged in containers that hold 12 ounces. If Jason buys one container, will he have enough mushrooms to make the sauce according to the recipe?

Check your answers on page 213.

Measuring Liquids

The chart below shows English and metric liquid measures.

ENGLISH		METRIC	
8 fluid ounces (fl oz)	= 1 cup (C)	1000 milliliters (mL)	= 1 liter (L)
2 cups	= 1 pint (pt)	100 centiliters (cL)	= 1 liter
16 fluid ounces	= 1 pint	10 deciliters (dL)	= 1 liter
2 pints	= 1 quart (qt)	10 liters	= 1 dekaliter (daL)
32 fluid ounces	= 1 quart	10 dekaliters	= 1 hectoliter (hL)
4 quarts	= 1 gallon (gal)	10 hectoliters	= 1 kiloliter (kL)
		1000 liters	= 1 kiloliter

Some soft drink and fruit juice containers show their contents in liters. Liters are also used in medicine and industry to make accurate measurements.

Remember, to change a large unit of measure to a smaller one, multiply; to change a small unit to a larger one, divide.

EXERCISE 13d

Change each unit of liquid measure to the one shown.

1. 5 pt = _____ fl oz
2. 4 L = _____ mL
3. 16 pt = _____ gal
4. 8 gal = _____ qt
5. 32 qt = _____ gal
6. 300 kL = _____ hL
7. 44 C = _____ pt
8. 20 daL = _____ L
9. 128 fl oz = _____ qt
10. 8 qt = _____ gal
11. 8 pt = _____ C
12. 5000 mL = _____ L

WORD PROBLEM

Answer the following question.

A doctor told Kerry to drink 16 fluid ounces of milk a day. Each day, Kerry wants to buy just the amount of milk he needs for that day. What size container of milk should he buy each day: a pint, a quart, or a gallon?

Check your answers on page 213.

Measuring Time

The list that follows shows the units that most people use to measure time. The words are different in different languages, but all nations use the same quantities when they communicate with each other.

60	seconds (sec)	=	1 minute (min)
60	minutes	=	1 hour (hr)
24	hours	=	1 day (da)
7	days	=	1 week (wk)
4	weeks	=	1 month (mo) [on average]
52	weeks	=	1 year (yr)
365	days	=	1 year
12	months	=	1 year

Units of time are changed into each other just as other units of measure are. Multiply to change a large unit into a smaller unit. Divide to change a small unit into a larger one.

EXERCISE 13e

Change each unit of time to the larger or smaller one shown.

1. 3 min = _____ sec
2. 42 da = _____ wk
3. 40 wk = _____ da
4. 240 sec = _____ min
5. 4 hr = _____ min
6. 780 wk = _____ yr
7. 730 da = _____ yr
8. 3 yr = _____ mo
9. 1 hr = _____ sec
10. 48 mo = _____ yr

WORD PROBLEM

Answer the following question.

Mildred has a 72-hour flu. How many days is she likely to be sick?

Check your answers on page 213.

Summary

Knowing how to change measurements into different units is useful in many situations. The following exercise will let you practice with the types of measures you have studied in this lesson.

Part A. Change each unit of measure to the one shown.

1. 2 mi = _____ ft
2. 216 in. = _____ yd
3. 70 daL = _____ L
4. 960 min = _____ hr
5. 15,840 ft = _____ mi
6. 50 m = _____ dam
7. 60 mo = _____ yr
8. 20 dg = _____ cg
9. 12 ft = _____ in.
10. 19,360 yd = _____ mi
11. 4 dag = _____ hg
12. 520 wk = _____ yr
13. 72 in. = _____ ft
14. 6 yd = _____ in.
15. 64 oz = _____ lb
16. 2 T = _____ lb
17. 52 wk = _____ da
18. 72 hr = _____ da
19. 256 oz = _____ lb
20. 16,000 lb = _____ T
21. 16 qt = _____ pt
22. 64 fl oz = _____ qt
23. 24 qt = _____ C
24. 8 L = _____ dL
25. 200 cg = _____ mg

Part B. Solve each problem.

1. Andrea knows that when she changes the oil in her airplane, she needs 4 gallons of oil. How many quarts of oil must she buy for an oil change?

2. Georgia made 16 pints of ice cream. She wants to pack the ice cream into quart containers. How many quart containers does she need?

3. Susan bought 13 yards of ribbon for presents she was wrapping. How many inches of ribbon did she buy?

4. Fred bought a roll of tape for his workshop. The tape was 864 inches long. How many yards of tape did he buy?

5. Alberto was sent away to a training program for his new job. He was at the training school for 7 weeks. How many days was he away?

Check your answers on page 214.

Working with Units of Measure:
The English System

In this lesson, you will learn to work with units of measure in the English system. You will practice adding, subtracting, multiplying, and dividing measurements. (Working with measurements in the metric system involves decimals, so you will study that in Unit 2.)

Adding Measurements

Adding units of measure is like adding other whole numbers. If a measurement includes two different units of measure, such as pounds and ounces, add each unit separately.

Example 1:   Add 3 pounds 4 ounces + 2 pounds 6 ounces.

Step 1	**Step 2**	**Step 3**
3 lb 4 oz	3 lb 4 oz	3 lb 4 oz
+ 2 lb 6 oz	+ 2 lb 6 oz	+ 2 lb 6 oz
	10 oz	5 lb 10 oz

STEP 1: Set up the problem with pounds under pounds and ounces under ounces.

STEP 2: Add the ounces: 4 + 6 = 10. Write the 10 under ounces.

STEP 3: Add the pounds: 3 + 2 = 5. Write the 5 under pounds. The answer is 5 pounds 10 ounces.

EXERCISE 14a

Add to solve the problems.

1. 6 yd 1 ft + 5 yd 1 ft =

2. 6 hr 30 min + 4 hr 20 min =

3. 4 T 800 lb + 2 T 800 lb =

4. 6 gal 1 qt + 4 gal 2 qt =

5. 30 min 10 sec + 16 min 8 sec =

6. 12 lb 4 oz + 9 lb 3 oz =

7. 26 mi 365 yd + 7 mi 77 yd =

8. 12 mo 2 wk + 8 mo 1 wk =

9. 12 T 314 lb + 9 T 95 lb =

10. 24 qt 1 pt + 37 qt =

11. 82 gal 1 qt + 19 gal 2 qt =

12. 6 qt 24 fl oz + 9 qt 7 fl oz =

Look for the key word that tells you to add to solve the following problem.

Harry worked out twice in one day preparing for a cross-country race. In the morning he ran for 35 minutes 38 seconds. That afternoon he ran for 24 minutes 11 seconds. What was the total amount of time Harry ran that day?

Check your answers on page 214.

Sometimes after you add measurements, some of the smaller units in the sum need to be changed to larger units. For instance, if a sum is 2 hours 65 minutes, it should be changed to 3 hours 5 minutes.

Example 2: Add 2 quarts 24 fluid ounces + 1 quart 16 fluid ounces.

Step 1	**Step 2**	**Step 3**
2 qt 24 fl oz	1 qt 8 fl oz	3 qt
+ 1 qt 16 fl oz	32$\overline{)40}$ fl oz	+ 1 qt 8 fl oz
3 qt 40 fl oz	32	4 qt 8 fl oz
	8	

STEP 1: Set up the problem. Add the fluid ounces: $24 + 16 = 40$. Add the quarts: $2 + 1 = 3$. The sum is 3 quarts 40 fluid ounces.

STEP 2: In the sum, 40 fluid ounces is more than 1 quart. To change some of the fluid ounces to quarts, divide by 32, the number of fluid ounces in a quart: $40 \div 32 = 1$ r8, or 1 quart 8 fluid ounces.

STEP 3: Add the changed fluid ounces (1 qt 8 fl oz) to the 3 quarts in the sum to get the final answer, 4 quarts 8 fluid ounces.

There is another way to do the same kind of problem. Before you add, change the large units in each measurement to smaller units.

Example 3: Add 2 quarts 24 fluid ounces + 1 quart 16 fluid ounces.

Step 1	**Step 2**	**Step 3**	**Step 4**
32	32	88 fl oz	4 qt 8 fl oz
× 2 qt	× 1 qt	48 fl oz	32$\overline{)136}$ fl oz
64 fl oz	32 fl oz	136 fl oz	128
+ 24 fl oz	+ 16 fl oz		8
88 fl oz	48 fl oz		

STEP 1: Multiply the 2 qt in the first measurement by 32 to change them to fluid ounces. Add the result, 64 fl oz, to the 24 fl oz that are already part of the measurement. The first measurement becomes 88 fl oz.

STEP 2: Multiply the 1 qt in the second measurement by 32 to change it to fluid ounces. Add the result, 32 fl oz, to the 16 fl oz that are already part of the measurement. The second measurement becomes 48 fl oz.

STEP 3: Add the two changed measurements: 88 + 48 = 136.

STEP 4: Divide by 32 to change 136 fluid ounces to quarts and fluid ounces.

EXERCISE 14b

Add to solve each of the following problems. Use either of the methods in Examples 2 and 3.

1. 15 min 45 sec + 20 min 20 sec =

2. 16 lb 10 oz + 20 lb 12 oz =

3. 5 qt 1 pt + 9 qt 1 pt =

4. 3 ft 8 in. + 6 ft 6 in. =

5. 65 lb 14 oz + 40 lb 12 oz =

6. 3 C 6 fl oz + 8 C 4 fl oz =

7. 12 wk 6 da + 9 wk 3 da =

8. 321 T 1840 lb + 99 T 605 lb =

9. 45 min 32 sec + 5 min 50 sec =

10. 6 yr 9 mo + 5 yr 4 mo =

11. 2 pt 1 C + 8 pt 1 C =

12. 12 min 42 sec + 9 min 51 sec =

WORD PROBLEM

Look for the key words that tell you to add to solve the following problem.

Charles worked two jobs to save enough money to buy a sports car. On one job he worked 40 hours and 30 minutes per week. On his second job he worked 15 hours and 30 minutes per week. How many hours per week did Charles work in all?

Check your answers on page 214.

Subtracting Measurements

Subtracting measurements is the same as subtracting whole numbers. You need to borrow from larger units of measure when there are not enough smaller units to subtract from.

Example 4: Find 15 pounds 4 ounces minus 8 pounds 9 ounces.

Step 1	Step 2	Step 3
	14 20	14 20
15 lb 4 oz	1̶5̶ lb 4̶ oz	1̶5̶ lb 4̶ oz
− 8 lb 9 oz	− 8 lb 9 oz	− 8 lb 9 oz
		6 lb 11 oz

STEP 1: Set up the problem. Be sure to write pounds under pounds and ounces under ounces.

STEP 2: Because you cannot subtract 9 ounces from 4 ounces, you need to borrow 1 pound from the pound column, leaving 14, and change it to ounces. Add it to the 4 ounces in the ounce column: 16 + 4 = 20.

STEP 3: Now subtract the two columns: 20 oz − 9 oz = 11 oz and 14 lb − 8 lb = 6 lb. The answer is 6 lb 11 oz.

As in addition, the other way to do the same kind of subtraction problem is to change the larger units in the problem to smaller units first.

Example 5: Find 15 pounds 4 ounces minus 8 pounds 9 ounces.

Step 1	Step 2	Step 3	Step 4
16	16	244 oz	6 lb 11 oz
× 15 lb	× 8 lb	− 137 oz	16) 107 oz
80	128 oz	107 oz	96
16	+ 9 oz		11
240 oz	137 oz		
+ 4 oz			
244 oz			

STEP 1: Multiply the 15 lb in the first measurement by 16 to change them to ounces. Add the product, 240 oz, to the 4 oz that are already part of the measurement. The first measurement becomes 244 oz.

STEP 2: Multiply the 8 lb in the second measurement by 16 to change them to ounces. Add the product, 128 oz, to the 9 oz that are already part of the measurement. The second measurement becomes 137 oz.

STEP 3: Subtract the two changed measurements: 244 − 137 = 107.

STEP 4: Divide by 16 to change 107 oz to pounds and ounces.

EXERCISE 14c

Subtract to solve each of the following problems. Use either of the methods in Examples 4 and 5.

1. 10 gal 3 qt − 7 gal 1 qt =

2. 15 wk − 6 wk 3 da =

3. 12 T 500 lb − 8 T 300 lb =

4. 8 yd 2 ft − 6 yd 1 ft =

5. 4 lb 4 oz − 2 lb 8 oz =

6. 30 min 10 sec − 18 min 25 sec =

7. 36 ft 9 in. − 20 ft 9 in. =

8. 10 qt 2 fl oz − 6 qt 4 fl oz =

9. 16 lb − 4 lb 6 oz =

10. 3 mi − 1500 yd =

11. 46 ft 10 in. − 8 ft 11 in. =

12. 55 min 14 sec − 16 min 20 sec =

WORD PROBLEM

Look for the key word that tells you to subtract to solve the following problem.

Terry bought a sack of onions that weighed 10 pounds 8 ounces. A recipe she was using called for 2 pounds 12 ounces of onions. If Terry followed the recipe, what was the weight of the onions left in the sack?

Check your answers on page 217.

Multiplying Measurements

Measurements are multiplied just as whole numbers are. When you multiply a measurement made of more than one unit of measure, multiply each unit separately.

Example 6: Multiply 3 quarts 4 fluid ounces by 5.

Step 1	Step 2	Step 3
3 qt 4 fl oz	3 qt 4 fl oz	3 qt 4 fl oz
× 5	× 5	× 5
	20 fl oz	15 qt 20 fl oz

STEP 1: Set up the problem with the measurement on top.

STEP 2: Multiply the fluid ounces: 4 fl oz × 5 = 20 fl oz.

STEP 3: Multiply the quarts: 3 qt × 5 = 15 qt. The answer is 15 qt 20 fl oz.

EXERCISE 14d

Multiply to solve each problem.

1. 3 T 200 lb × 8 =

2. 4 da 6 hr × 3 =

3. 6 ft 2 in. × 5 =

4. 3 qt 2 fl oz × 12 =

5. 11 lb 2 oz × 7 =

6. 3 mi 220 yd × 3 =

7. 16 yd 1 ft × 2 =   8. 42 mi 316 ft × 10 =   9. 37 min 11 sec × 5 =

10. 14 lb 1 oz × 12 =   11. 8 qt 4 fl oz × 3 =   12. 9 T 147 lb × 6 =

WORD PROBLEM

Look for the key words that tell you to multiply to solve the following problem.

Alicia was buying nylon line to make docking lines for her boat. She needed 6 docking lines. Each line had to be 14 feet 1 inch long. What was the total length of nylon line she needed?

Check your answers on page 219.

Example 2 in this lesson, on page 86, showed you that when you add two measurements you sometimes have to change some of the smaller units in the sum to larger units. The same kind of change is sometimes necessary in products you get when you multiply a measurement.

Example 7:   Multiply 4 yards 2 feet by 2.

Step 1	Step 2	Step 3
4 yd 2 ft	1 yd 1 ft	8 yd
× 2	3)4 ft	+ 1 yd 1 ft
8 yd 4 ft		9 yd 1 ft

STEP 1: Set up the problem. Multiply the feet: 2 ft × 2 = 4 ft. Multiply the yards: 4 yd × 2 = 8 yd.

STEP 2: Divide to change some of the feet to yards: 4 ft ÷ 3 = 1 yd 1 ft.

STEP 3: Add in the yards from the product to find the final answer.

As in addition and subtraction, the other way to do the same kind of multiplication problem is to change the larger units in the problem to smaller units first.

Example 8:   Multiply 4 yards 2 feet by 2.

Step 1	Step 2	Step 3
4 yd	14 ft	9 yd 1 ft
× 3	× 2	3)28 ft
12 ft	28 ft	
+ 2 ft		
14 ft		

STEP 1: Multiply the 4 yd in the measurement by 3 to change them to feet. Add the product, 12 ft, to the 2 ft that are already part of the measurement. The measurement becomes 14 ft.

---

90   UNIT 1: Whole Numbers

STEP 2: Multiply the changed measurement by 2: $14 \times 2 = 28$.

STEP 3: Divide by 3 to change 28 ft to yards and feet.

Multiply to solve the following problems. Use either of the methods in Examples 7 and 8.

1. 1 mi 440 yds $\times$ 4 =

2. 4 lb 12 oz $\times$ 8 =

3. 14 pt 8 oz $\times$ 6 =

4. 72 min 23 sec $\times$ 4 =

5. 12 qt 1 pt $\times$ 23 =

6. 4 yr 8 mo $\times$ 9 =

7. 32 wk 5 da $\times$ 4 =

8. 16 C 4 fl oz $\times$ 21 =

9. 7 yd 28 in. $\times$ 4 =

10. 3 hr 24 min $\times$ 3 =

11. 24 mi 500 yd $\times$ 32 =

12. 32 gal 3 qt $\times$ 7 =

WORD PROBLEM

Look for the key words that tell you to multiply to solve the following problem.

George worked 44 hours 30 minutes each week for 7 weeks to help out with a special project. How many hours and minutes did George work in all during those 7 weeks?

Check your answers on page 219.

Dividing Measurements

To divide measurements, first divide the larger unit, and then divide the smaller unit.

Example 9: Divide 9 gallons 3 quarts by 3.

Step 1	Step 2
$\dfrac{3 \text{ gal}}{3 \overline{)9 \text{ gal } 3 \text{ qt}}}$	$\dfrac{3 \text{ gal } 1 \text{ qt}}{3 \overline{)9 \text{ gal } 3 \text{ qt}}}$

STEP 1: Divide the gallons: $9 \div 3 = 3$. Write the 3 above the gallons.

STEP 2: Divide the quarts: $3 \div + = 1$. Write the 1 above the quarts. The answer is 3 gallons 1 quart.

Divide to solve the following problems.

1. 21 lb 14 oz ÷ 7 =   2. 21 wk 6 da ÷ 3 =   3. 48 yd 16 in. ÷ 8 =

4. 20 gal 15 fl oz ÷ 5 =   5. 28 lb 12 oz ÷ 4 =   6. 66 ft 11 in. ÷ 11 =

7. 42 mo 2 wk ÷ 2 =   8. 75 hr 25 min ÷ 5 =   9. 24 ft 9 in. ÷ 3 =

10. 81 T 162 lb ÷ 27 =   11. 164 da 82 min ÷ 41 =   12. 114 gal 3 qt ÷ 3 =

WORD PROBLEM

Look for the key words that tell you to divide to solve the following problem.

Francesca needed to cut a board 18 feet 9 inches long into 3 equal lengths. How many feet long was each board after she made the cuts?

Check your answers on page 222.

When the units in a measurement will not divide evenly, it is best to change the larger units in the problem to smaller units before dividing.

Example 10:   Divide 12 feet 6 inches by 5.

Step 1	Step 2	Step 3
12 ft	30 in.	2 ft 6 in.
× 12	5) 150 in.	12) 30 in.
24	15	24
12	00	6
144 in.		
+ 6 in.		
150 in.		

STEP 1: Multiply the 12 ft in the measurement by 12 to change them to inches. Add the product, 144 in., to the 6 in. that are already part of the measurement. The measurement becomes 150 in.

STEP 2: Divide the changed measurement by 5: 150 ÷ 5 = 30.

STEP 3: Divide by 12 to change 30 in. to feet and inches: 30 ÷ 12 = 2 with a remainder of 6. The answer is 2 ft 6 in.

Divide to solve the following problems. Change the larger units to smaller ones first.

1. 9 ft 4 in. ÷ 5 =

2. 17 wk 1 da ÷ 2 =

3. 11 pt 4 fl oz ÷ 10 =

4. 16 T 500 lb ÷ 5 =

5. 22 gal 2 qt ÷ 6 =

6. 14 lb 6 oz ÷ 5 =

7. + mi 181 yd ÷ 3 =

8. 23 mo 12 da ÷ 9 =

9. 10 pt 1 C ÷ 3 =

10. 9 mi 281 yd ÷ 7 =

11. 15 C 4 fl oz ÷ 4 =

12. 10 yr 8 mo ÷ 8 =

WORD PROBLEM

Look for the key words that tell you to divide to solve the following problem.

Sam ran the same distance each day for 3 days. He ran a total of 22 miles 232 yards. How far did he run each day?

Check your answers on page 222.

Lesson 15

Finding Perimeter and Area

Triangles, **squares**, and **rectangles** are closed figures with straight sides.

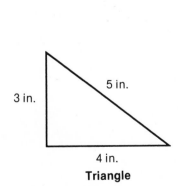

3 in. 5 in. 4 in.

Triangle

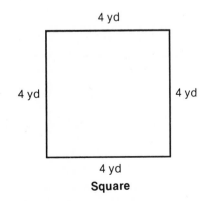

4 yd 4 yd 4 yd 4 yd

Square

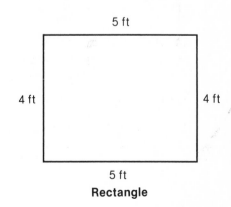

5 ft 4 ft 4 ft 5 ft

Rectangle

Triangles have three sides. Squares and rectangles have four sides. All the sides of a square are the same length. The opposite sides of a rectangle are the same length.

Perimeter: Triangles, Squares, and Rectangles

The distance around a straight-sided figure is called its **perimeter**. To find the perimeter of a figure, add the lengths of all its sides.

Example 1: To buy enough fencing to go around this rectangular parking lot, you would need to know its perimeter. Find the perimeter of the parking lot.

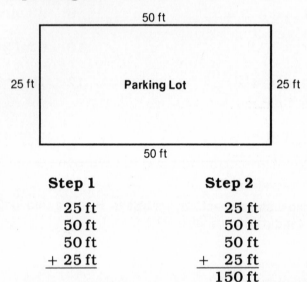

Step 1	Step 2
25 ft	25 ft
50 ft	50 ft
50 ft	50 ft
+ 25 ft	+ 25 ft
	150 ft

STEP 1: Write the lengths of all the sides of the parking lot in a column for addition. (Note: The order in which you write the lengths of the sides does not matter.)

STEP 2: Add. The perimeter of the parking lot is 150 feet.

Since all four sides of a square are the same length, it is faster to multiply to find its perimeter.

Example 2: Find the perimeter of this square.

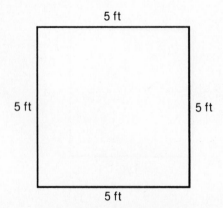

	Step 1	**Step 2**

Step 1   **Step 2**

$$\begin{array}{r} 5\text{ ft} \\ \times\ 4 \\ \hline \end{array} \qquad \begin{array}{r} 5\text{ ft} \\ \times\ 4 \\ \hline 20\text{ ft} \end{array}$$

STEP 1: Set up the problem so that you can multiply 5 feet, the length of one side of the square, by 4, the number of sides.

STEP 2: Multiply. The perimeter of the square is 20 feet.

When all three sides of a triangle are the same length, you can multiply by 3 to find its perimeter. It is faster than adding.

EXERCISE 15a

Part A. Find the perimeter of each of these figures.

1.
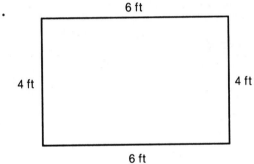

2.
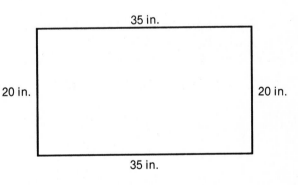

3.
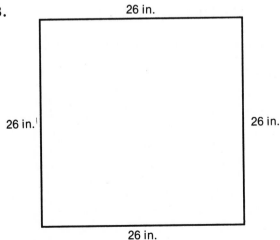

4.
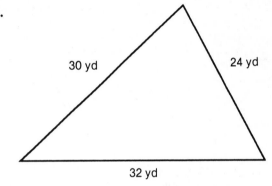

5.

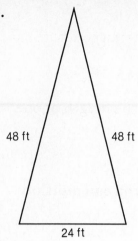

6.

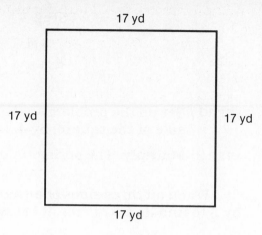

7.

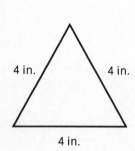

8.

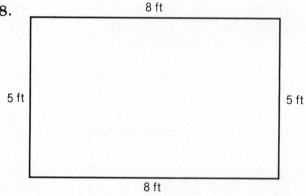

Part B. The lengths of the sides of eight different figures are given below. Tell what kind of figure each one is and find its perimeter.

1. 24 ft, 12 ft, 24 ft, 12 ft

2. 7 yd, 7 yd, 7 yd, 7 yd

3. 17 in., 13 in., 19 in.

4. 43 ft, 29 ft, 43 ft, 29 ft

5. 12 ft, 12 ft, 12 ft

6. 33 in., 10 in., 33 in., 10 in.

7. 15 in., 11 in., 18 in.

8. 9 in., 9 in., 9 in.

WORD PROBLEM

Solve the following problem.

Harold Brown decided to enclose part of his back yard for a dog run for his two dogs. He settled on a rectangle whose sides would measure 32 feet, 12 feet, 32 feet, and 12 feet. How many feet of fencing does he need to make the dog run?

Check your answers on page 224.

Area: Rectangles and Squares

The amount of space inside a figure is called its **area**. Area is described in square units of measure.

The space inside this rectangle is divided into 12 equal parts. Each of the parts is a square with sides 1 foot long.

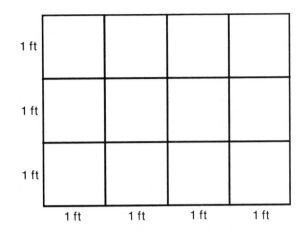

These square parts are the units used to measure area. The area of a figure is the number of square units inside it. The rectangle has an area of 12 square feet (sq ft).

If the sides of the squares inside the rectangle measured 1 inch, the area of the rectangle would be 12 square inches (sq in.). If they measured 1 yard, the area would be 12 square yards (sq yd).

One way to find the area of a figure is to count the square units inside it.

Example 3: Find the area of this rectangle.

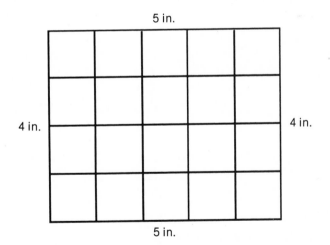

METHOD: Count the number of squares that cover the rectangle. There are 20 squares. Each square has 1-inch sides. Therefore, the area of the rectangle is 20 square inches (20 sq in.).

A faster way to find the area of a four-sided figure (a square or a rectangle) is to multiply its **length** by its **width**. This method can be expressed in a formula:

$$\text{Area} = \text{length} \times \text{width}$$

Example 4: Find the area of this rectangle.

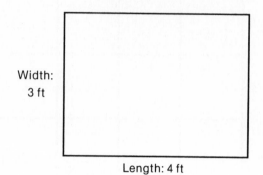

Width: 3 ft

Length: 4 ft

Step 1

Area = length × width

Step 2

Area = length × width
Area = 4 ft × 3 ft

Step 3

Area = length × width
Area = 4 ft × 3 ft
Area = 12 sq ft

STEP 1: Write the formula for the area of a four-sided figure.

STEP 2: Rewrite the formula, substituting *4 ft* and *3 ft* for the words *length* and *width*.

STEP 3: Multiply 4 ft by 3 ft. Rewrite the formula to state the area, 12 sq ft. (Remember that area is expressed in square units of measure.)

EXERCISE 15b

Use the formula for the area of a four-sided figure to find the area of each of the following figures.

1.

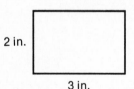

2 in.

3 in.

2.

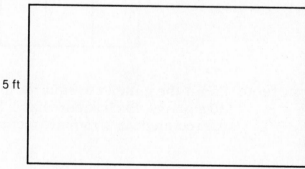

5 ft

9 ft

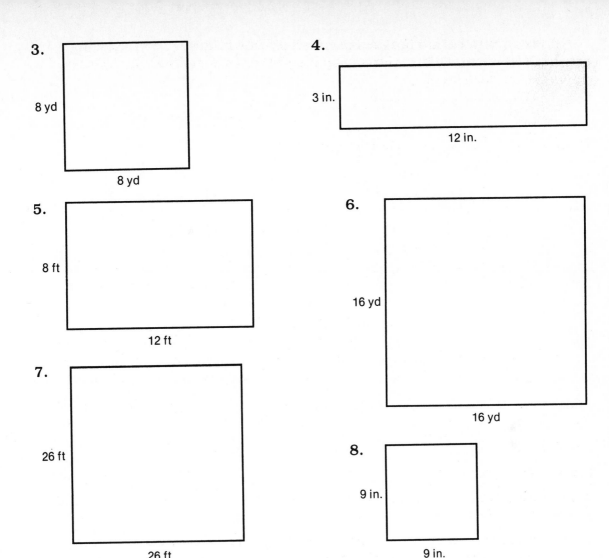

3.

8 yd

8 yd

4.

3 in.

12 in.

5.

8 ft

12 ft

6.

16 yd

16 yd

7.

26 ft

26 ft

8.

9 in.

9 in.

WORD PROBLEM

Solve the following problem.

Linda wants to lay 1-foot square tiles in her kitchen. Her kitchen is 12 feet long and 10 feet wide. How many tiles will she need to buy?

Check your answers on page 224.

Area: Triangles

Finding the area of a triangle is similar to finding the are of a rectangle or of a square. To find the area of a triangle, multiply its **base** by its **height** and divide by 2. This can be expressed in a formula:

$$\text{Area} = \frac{\text{base} \times \text{height}}{2}$$

You can think of a triangle as half a four-sided figure. The following drawings illustrate this.

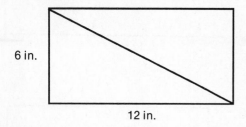

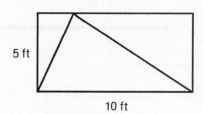

The base of the triangle on the left is 12 inches long. Its height is 6 inches. Its area is half the area of a rectangle that measures 12 inches by 6 inches. The base of the triangle on the right is 10 feet. Its height is 5 feet. Its area is half the area of a rectangle that measures 10 feet by 5 feet.

Example 5: The areas of these two triangles are the same. Use the formula for the area of a triangle to find out what those areas are.

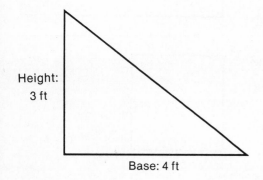

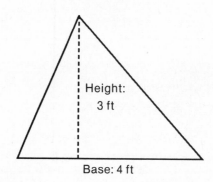

Step 1

$$\text{Area} = \frac{\text{base} \times \text{height}}{2}$$

Step 2

$$\text{Area} = \frac{\text{base} \times \text{height}}{2}$$

$$\text{Area} = \frac{4 \text{ ft} \times 3 \text{ ft}}{2}$$

Step 3

$$\text{Area} = \frac{\text{base} \times \text{height}}{2}$$

$$\text{Area} = \frac{4 \text{ ft} \times 3 \text{ ft}}{2}$$

$$\text{Area} = 6 \text{ sq ft}$$

STEP 1: Write the formula for the area of a triangle.

STEP 2: Rewrite the formula, substituting *4 ft* and *3 ft* for the words *base* and *height*.

STEP 3: Multiply 4 ft by 3 ft and divide the result by 2. Rewrite the formula to state the area, 6 sq ft (Remember that area is expressed in square units of measure.)

Use the formula for the area of a triangle to find the area of each of the following figures.

1.

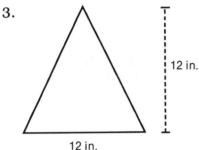

6 in.

9 in.

2.

4 ft

12 ft

3.

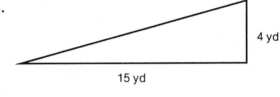

12 in.

12 in.

4.

4 yd

15 yd

WORD PROBLEM

Solve the following problem.

Malcolm wants to know the area of the triangular counter he has in his shop. The counter has a base of 8 feet and a height of 4 feet. What is the area of the counter?

Check your answers on page 224.

Lesson 16

Working with Perimeter and Area

Perimeter: Measurements in Different Units

Some problems that ask you to find the perimeter of a figure give measurements in different units. For example, the length may be given in feet, and the width in inches or yards. To solve this kind of problem, you first need to change one or more measurements so that they are all in the same units. It is usually best to change all measurements to the smaller unit. It is also usually best to give the answer in the larger unit.

The following example shows how to solve such problems. Skills you studied in Lesson 13 are used in Steps 1, 2, and 4. A skill you studied in Lesson 15 is used in Step 3.

Example 1: Find the perimeter of this rectangle.

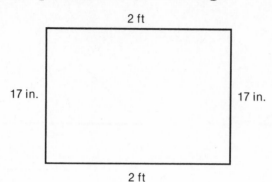

Step 1	**Step 2**	**Step 3**	**Step 4**
1 ft = 12 in.	12	24 in.	6 ft 10 in.
	× 2 ft	24 in.	12⟌82 in.
	24 in.	17 in.	72
		+ 17 in.	10
		82 in.	

STEP 1: To change the length, the measurement that is given in feet, to inches, first find the number of inches in 1 ft. There are 12 in. in 1 ft.

STEP 2: Multiply the length by 12: 2 ft × 12 = 24 in. This is the length in inches.

STEP 3: To find the perimeter of the rectangle, add the measurements of the four sides in inches. The perimeter is 82 in.

STEP 4: Because the original measurements were given in feet and inches, change the answer to feet with a remainder of inches. Divide by 12, the number of inches in a foot. The perimeter is 6 ft 10 in.

EXERCISE 16a

Find the perimeter of each of these figures.

1.

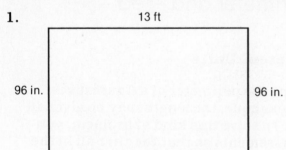

2.

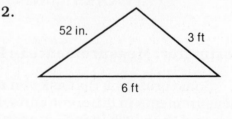

3.

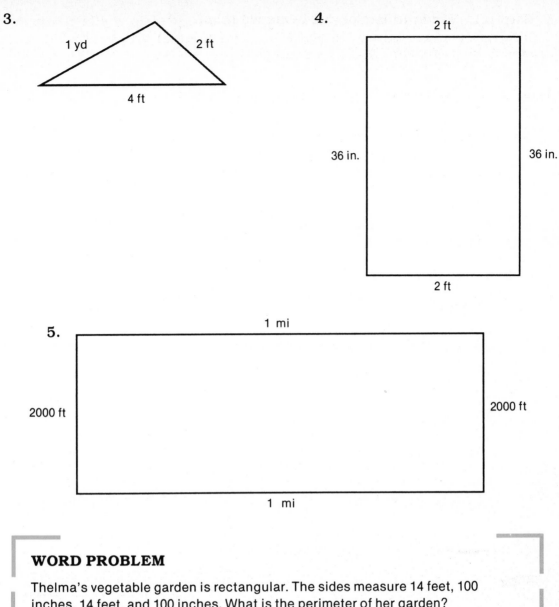

1 yd   2 ft

4 ft

4.

2 ft

36 in.   36 in.

2 ft

5.

1 mi

2000 ft   2000 ft

1 mi

WORD PROBLEM

Thelma's vegetable garden is rectangular. The sides measure 14 feet, 100 inches, 14 feet, and 100 inches. What is the perimeter of her garden?

Check your answers on page 224.

Area: L-shaped and T-shaped Figures

An L-shaped figure or a T-shaped figure is nothing more than two rectangles joined together. To find the area of such a figure first find the area of each of the two rectangles. Then add the two areas.

The following example shows how to find the area of such figures. Skills you studied in Lesson 15 are used in the example.

Example 2: Find the area of this L-shaped room.

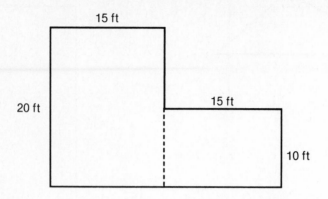

Step 1	Step 2	Step 3
Area = length × width	Area = length × width	300 sq ft
Area = 15 ft × 20 ft	Area = 15 ft × 10 ft	+ 150 sq ft
Area = 300 sq ft	Area = 150 sq ft	450 sq ft

STEP 1: Find the area of one of the rectangles.

STEP 2: Find the area of the other rectangle.

STEP 3: Add the areas of the two rectangles. The area of the L-shaped room is 450 sq ft.

EXERCISE 16b

Find the area of each of the following figures.

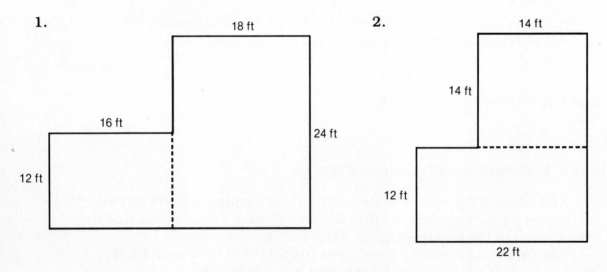

1.

2.

3.

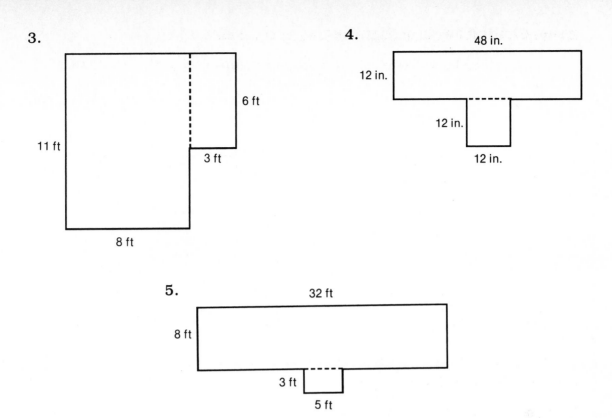

4. 48 in. 12 in. 12 in. 12 in.

5. 32 ft 8 ft 3 ft 5 ft

WORD PROBLEM

Solve the following problem.

This is a sketch of Pat's L-shaped metal shop. What is the area of the shop?

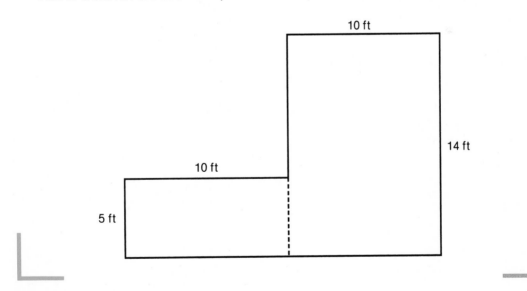

Check your answers on page 225.

Area: Shaded Parts of Figures

Figures can be made up of a rectangle within a rectangle or a square within a square. Some problems ask you to find the area of part of such a figure. The following example shows how to solve such problems. Skills you studied in Lesson 15 are used in the example.

Example 3: Find the area of the shaded part of the following figure.

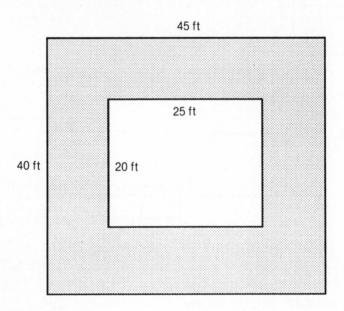

Step 1	Step 2	Step 3
Area = length × width	Area = length × width	1800 sq ft
Area = 45 ft × 40 ft	Area = 25 ft × 20 ft	− 500 sq ft
Area = 1800 sq ft	Area = 500 sq ft	1300 sq ft

STEP 1: Find the area of the larger rectangle.

STEP 2: Find the area of the smaller rectangle.

STEP 3: Subtract the area of the smaller rectangle from the area of the larger rectangle. The result is the area of the shaded part of the figure, 1300 sq ft.

Find the area of the shaded part of each of the following figures.

1.

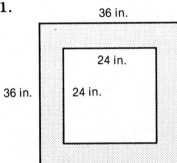

2.

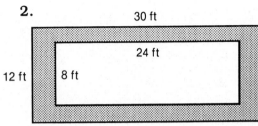

3.

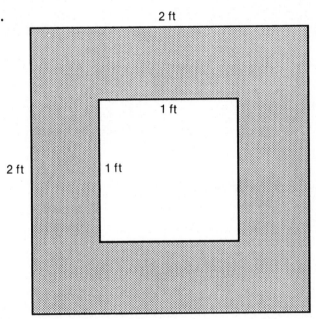

4.

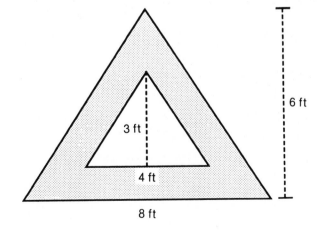

5.

Solve the following problem.

This is a drawing of one city block. The shaded part is the sidewalk that runs around the outside of the block. What is the area of the sidewalk?

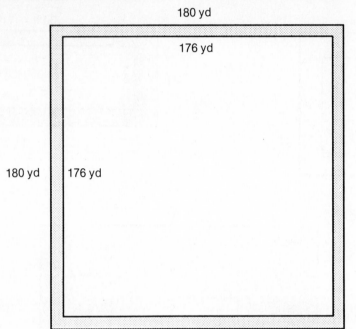

Check your answers on page 226.

Lesson
17

Finding Averages

You probably often hear and use the word **average**. You might hear it used in these ways:

The average high temperature for this date is 76 degrees Farenheit.
Helaine's bowling average is 144 pins.
The average score on the test was 83 percent.
The plant has an average of 212 people working each shift.

To find the average of a set of numbers, divide the sum of the set of numbers by the number of numbers in the set.

Example: Find the average of this set of numbers: 85, 98, 84, 75, and 93.

Step 1	Step 2

$$
\begin{array}{r}
85 \\
98 \\
84 \\
75 \\
+\ 93 \\
\hline
435
\end{array}
\qquad
\begin{array}{r}
87 \\
5\overline{)435} \\
\underline{40} \\
35 \\
\underline{35} \\
0
\end{array}
$$

STEP 1: Add to find the sum of the five numbers in the set.

STEP 2: Divide the sum by 5, the number of numbers in the set. The average of this set of numbers is 87.

EXERCISE 17

Find the average of each set of numbers.

1. 12, 15, 25, 8

2. $231, $321, $132

3. 123, 56, 72, 96, 65, 54, 45

4. 41, 42, 43, 44, 45, 46

5. 700, 650, 900, 850, 930

6. 48, 52, 48, 39, 57, 50

7. 54, 80, 40, 48, 78

8. $22, $30, $18, $21, $34, $307

9. 24 in., 120 in., 228 in., 144 in., 84 in., 177 in.

10. 236 mi, 52 mi, 18 mi, 558 mi, 241 mi, 1000 mi, 14 mi, 41 mi, 650 mi, 650 mi

11. 311, 642, 213, 345, 744

12. 4, 6, 24, 25, 27, 34

13. 9786, 2435, 9001, 3142

14. 90, 7, 1, 34, 18, 19, 33, 46

15. $14, $64, $72, $87, $108

16. 2 oz, 4 oz, 8 oz, 16 oz, 32 oz, 64 oz

WORD PROBLEM

Solve the following problem.

In the first 5 weeks on his new job, Raul's paychecks were $325, $295, $330, $350, and $275. What was his average pay per week.

Check your answers on page 226.

Solving Whole-Number Word Problems

Here are the steps you should take to solve word problems.
- Understand the question.
- Find the facts you need to solve the problem.
- Decide which math operation to use.
- Estimate the solution.
- Solve the problem and check your answer.

Understanding the Question

Most of the problems on the GED are word problems. They test how you apply your arithmetic skills to problem-solving situations.

The first step in solving a word problem is to read the problem carefully. If there are words that you do not understand, try to figure out what they mean by the way they are used in the problem. Then reread the problem and pay close attention to the sentence that contains the question. It may help you decide how to solve the problem if you rewrite the question.

Example 1: A voter registration drive enrolled 2550 people in 1984, 1437 people in 1985, 880 in 1986, and 643 in 1987. Altogether, how many voters were signed up during these four years?

Step 1	Step 2	Step 3
Enrolled means "were signed up."	Altogether, how many voters were signed up during these four years?	Altogether, how many voters were enrolled in 1984, 1985, 1986, and 1987?

STEP 1: Read the problem carefully. If you're not sure of the meaning of a word, look for context clues to its meaning. In this problem, the words *enrolled* and *were signed up* are used to describe the same activity.

STEP 2: Reread the problem. Especially notice the sentence that contains the question.

STEP 3: Rewrite the question if doing so will help you decide how to solve the problem.

Read each problem carefully. Then choose the best rewrite of the question in the problem.

1. Five people shared the price of a birthday dinner. If the bill was $75, how much did each person pay?

   (1) How many people are at the birthday dinner?
   (2) What is the cost of 5 dinners?
   (3) What is each person's share of the bill?
   (4) If 1 dinner costs $75, what is the cost of 5 dinners?
   (5) How much change did the people receive?

2. At the first bus stop, 7 people got on the bus. At the second stop, 4 people got on. At the next stop, 3 people got on. At the fourth stop, 7 people got on. Altogether, how many people were on the bus after the fourth stop?

   (1) After which stop were there 14 people on the bus?
   (2) How many stops did the bus make?
   (3) How many people got on the bus at the third stop?
   (4) After the fourth stop, how many people were on the bus?
   (5) How much did each person pay in bus fare?

3. Mario operates his metal-stamping machine 90 times per hour. How many times does Mario operate his machine in a 7-hour workday?

   (1) How many pieces of metal does Mario stamp each hour?
   (2) How many times does Mario operate his machine in 7 hours?
   (3) How long does it take Mario to operate the machine 90 times?
   (4) How much is Mario paid for a 7-hour workday?
   (5) How many hours does Mario work each day?

4. In November, 395 people worked at the post office. In December 540 people worked at the post office. How many more people worked at the post office in December than in November?

   (1) How many people worked at the post office in November and December?
   (2) How many people worked at the post office this year?
   (3) How many people worked at the post office in October?
   (4) Compared with November, how many more people worked at the post office in December?
   (5) How many people did the post office employ in December?

5. Fifty-six people are working in the tomato fields. Each person picks 25 bushels of tomatoes a day. What is the total number of bushels picked in one day?

   (1) How many bushels of tomatoes do 56 people pick in one day if each one picks 25 bushels?
   (2) What is the price of a bushel of tomatoes?
   (3) How many hours do the 56 people work in one day?
   (4) How many bushels does each person pick in a day?
   (5) How much are tomato pickers paid for picking 25 bushels?

Check your answers on page 227.

Finding the Facts

When you know what question a problem asks you to answer, you must find the facts you need to solve the problem. The facts in a word problem are the numbers you will add, subtract, multiply, or divide.

Example 2: A voter registration drive enrolled 2550 people in 1984, 1437 people in 1985, 880 in 1986, and 643 in 1987. Altogether, how many voters were signed up during these four years?

Step 1	**Step 2**
A voter registration drive enrolled **2550** people in **1984**, **1437** people in **1985**, **880** in **1986**, and **643** in **1987**.	A voter registration drive enrolled **2550** people in 1984, **1437** people in 1985, **880** in 1986, and **643** in 1987.

STEP 1: There are two types of numbers in this problem: years and the number of voters enrolled each year.

STEP 2: Reread the question. You must find *how many voters*. Therefore, the important numbers in the problem are the numbers of voters enrolled each year.

EXERCISE 18b

Read each problem carefully. Then choose the answer that contains the facts you need to solve the problem.

1. Five people shared the price of a birthday dinner. If the bill was $75, how much did each person pay?
   - (1) 5 dinners; $75 per dinner
   - (2) 5 people; 75 bills
   - (3) 75 people; $5 per dinner
   - (4) 5 people; a bill for $75
   - (5) 5 people; a bill for $15

2. At the first stop, 7 people got on the bus. At the second stop, 4 people got on. At the next stop, 3 people got on. At the fourth stop, 7 people got on. Altogether, how many people were on the bus after the fourth stop?
   - (1) 7 people; 4 people; 3 people; 7 people
   - (2) 7 people; 1 stop; 4 people; 2 stops
   - (3) 1 stop; 2 stops; 3 stops; 4 stops
   - (4) 3 stops; 3 people; 4 stops; 7 people
   - (5) 10 stops; 21 people

3. Mario operates his metal-stamping machine 90 times per hour. How many times does Mario operate his machine in a 7-hour work day?

    (1) 7 times per hour; 90 hours
    (2) 7 hours; 90 hours
    (3) 90 times per hour; 7 times per day
    (4) 90 times per day; 7 days per week
    (5) 90 times per hour; 7 hours per day

4. In November, 395 people worked at the post office. In December, 540 people worked at the post office. How many more people worked at the post office in December than in November?

    (1) 395 people in December; 540 people in November
    (2) 540 people in October; 395 people in November
    (3) 395 people in November; 540 people in December
    (4) 540 people in December; 395 people in January
    (5) 395 people at the post office; 540 people not at the post office

5. Fifty-six people are working in the tomato fields. Each person picks 25 bushels of tomatoes a day. What is the total number of bushels picked in one day?

    (1) 56 tomato fields; 25 people
    (2) 56 people; $25 per day
    (3) 25 people; 56 bushels of tomatoes
    (4) 56 people; 25 bushels per person
    (5) 25 bushels; $56 per bushel

Check your answers on page 227.

Deciding on the Operation

After you know what question the problem is asking you to answer and which facts you need to work with, you must decide what arithmetic operation to use. Ask yourself, "Do I need to add, subtract, multiply, or divide?" You know that certain key words, or combinations of key words, can help you to identify the operation you should use to solve a word problem. Here is a list of key words. Some of the words you already know, some are new.

Addition	Subtraction
altogether	decreased by
combined	difference
in all	fewer
increased by	how much less, less than
sum	how many more, more than
total	left
	remain

Multiplication	Division
altogether	divide; divide evenly
at	each
in all	equal
rate	share
times	
total	

Example 3: A voter registration drive enrolled 2550 people in 1984, 1437 people in 1985, 880 in 1986, and 643 in 1987. Altogether, how many voters were signed up during these four years?

Step 1	**Step 2**
Altogether, how many voters were signed up in these four years?	2550 voters
	1437 voters
	880 voters
	+ 643 voters

STEP 1: Reread the problem looking for key words. In this problem, the word *altogether* tells you to add. Think: Does adding make sense? Adding means finding the sum of a set of numbers. It makes sense.

STEP 2: Set up the addition problem using the facts you need to solve the problem.

EXERCISE 18c

Read each problem carefully. Look for the key words in the problem. Tell which operation you need to do to solve the problem.

1. Five people shared the price of a birthday dinner. If the bill was $75, how much did each person pay?

   (1) addition   (3) multiplication
   (2) subtraction   (4) division

2. At the first stop, 7 people got on the bus. At the second stop, 4 people got on. At the next stop, 3 people got on. At the fourth stop, 7 people got on. Altogether, how many people were on the bus after the fourth stop?

   (1) addition   (3) multiplication
   (2) subtraction   (4) division

3. Mario operates his metal-stamping machine 90 times per hour. How many times does Mario operate his machine in a 7-hour work day?

   (1) addition   (3) multiplication
   (2) subtraction   (4) division

4. In November, 395 people worked at the post office. In December, 540 people worked at the post office. How many more people worked at the post office in December than in November?

(1) addition   (3) multiplication
(2) subtraction   (4) division

5. Fifty-six people are working in the tomato fields. Each person picks 25 bushels of tomatoes a day. What is the total number of bushels picked in one day?

(1) addition   (3) multiplication
(2) subtraction   (4) division

Check your answers on page 227.

Estimating the Solution

Once you have found the facts you need to solve a problem and have decided which math operation to use, estimate the answer to the problem. Having an estimate of the answer to a problem will help you decide whether your final solution is correct.

To estimate the answer to a problem, round the facts in the problem and perform the required math operation.

Example 4:   A voter registration drive enrolled 2550 people in 1984, 1437 people in 1985, 880 in 1986, and 643 in 1987. Altogether, how many voters were signed up during these four years?

Step 1	Step 2	Step 3
2550 voters	2600 voters	2600 voters
1437 voters	1400 voters	1400 voters
880 voters	900 voters	900 voters
+ 643 voters	+ 600 voters	+ 600 voters
		5500 voters

STEP 1: Begin with the addition problem you set up using the facts from the problem.

STEP 2: Round the facts from the problem. (In this case, the facts are rounded to the nearest hundred. In each case, you need to decide how to round the facts.)

STEP 3: Using the rounded facts, add. The estimate of the correct answer is 5500 voters.

Read each problem carefully. Estimate the answer to the problem.

1. Five people shared the price of a birthday dinner. If the bill was $75, how much did each person pay?

2. At the first stop, 7 people got on the bus. At the second stop, 4 people got on. At the next stop, 3 people got on. At the fourth stop, 7 people got on. Altogether, how many people were on the bus after the fourth stop?

3. Mario operates his metal-stamping machine 90 times per hour. How many times does Mario operate his machine in a 7-hour work day?

4. In November, 395 people worked at the post office. In December, 540 people worked at the post office. How many more people worked at the post office in December than in November?

5. Fifty-six people are working in the tomato fields. Each person picks 25 bushels of tomatoes a day. What is the total number of bushels picked in one day?

Check your answers on page 227.

Solving the Problem and Checking Your Answer

In earlier lessons, you performed and checked the operations of addition, subtraction, multiplication, and division. The following word problems give you a chance to use those operations to solve problems.

When you have solved a problem, compare it with your estimate of the solution. If your solution is correct, it should be close to your estimate.

Part A. Solve and check these problems.

1. Five people shared the price of a birthday dinner. If the bill was $75, how much did each person pay?

2. At the first stop, 7 people got on the bus. At the second stop, 4 people got on. At the next stop, 3 people got on. At the fourth stop, 7 people got on. Altogether, how many people were on the bus after the fourth stop?

3. Mario operates his metal-stamping machine 90 times per hour. How many times does Mario operate his machine in a 7-hour work day?

4. In November, 395 people worked at the post office. In December, 540 people worked at the post office. How many more people worked at the post office in December than in November?

5. Fifty-six people are working in the tomato fields. Each person picks 25 bushels of tomatoes a day. What is the total number of bushels picked in one day?

Part B. Read each problem carefully, making sure you understand the question. Find the facts you need to solve the problem and decide which operation to use. Estimate the solution, and then solve the problem. Check your answer.

1. Last year, 26,489 cars came off the assembly line at a motor plant. This year 31,500 cars came off the assembly line. How many more cars came off the assembly line this year than last?

2. Jean drives a 12-mile messenger route 6 times per day. How many miles does she drive altogether on her route each day?

3. There are 375 people on a subway train. The train has 15 cars. If an equal number of people are in each car, how many people are in each car?

4. Carmine has a temporary job that pays $375 per week. How much will he earn if he works for 18 weeks?

5. On Friday night, 3576 people attended the concert. On Saturday night, attendance increased by 845 people. How many people went to Saturday night's concert?

6. Joy's dairy farm has 6279 cows. Each day an equal number of cows graze in each of the farm's 3 pastures. How many cows graze in each pasture?

7. Larissa earns $22,320 a year. Gus, her husband, has a job that pays $17,500. Their son Augie earns $240 each year, and their daughter Andrea earns $60 a year. What is the total annual income for the family?

8. Wallie enrolled in an art class. Each time he attends class, he pays a fee of $12. Wallie hopes to get to 20 classes. If he makes all 20, how much will he have paid in fees?

9. Joe wants to buy a tool that usually sells for $43. This week it is on sale for $37. If Joe buys the tool this week, how much money will he save?

10. When the Andreano family moved to town, there were 5408 people living there. Now, 10 years later, the population has grown to 9457. Compared with 10 years ago, how many more people are living in the town now?

Check your answers on page 227.

Solving Multistep Whole-Number
Word Problems

To solve many word problems, you need to perform more than one operation. You may need to add first and then multiply, for example. Often you need to add, subtract, multiply, or divide just to find a fact (a number) you need before you can solve a problem.

Example 1: Judd bought a box of screws for $3 and a screwdriver for $4. He gave the cashier a $10 bill. How much change did he receive?

Step 1	Step 2
$4 (screwdriver)	$10 (amount tendered)
+ 3 (screws)	− 7 (total cost)
$7 (total cost)	$ 3 (change)

STEP 1: To find how much change Judd received, you first need to find the cost of his purchases. Add the two prices to find the total cost.

STEP 2: To find out how much change Judd got, subtract the total cost from the amount he handed the cashier. He received $3 in change.

EXERCISE 19

Part A. Read each problem carefully. Choose the answer that tells which operations you need to do to solve each problem.

1. Louise works at a part-time job for 3 hours each day. She earns $5 per hour. How much money does she earn in 5 days?
   - (1) divide then multiply
   - (2) multiply then divide
   - (3) multiply then multiply
   - (4) add then multiply
   - (5) multiply then subtract

2. Jonah bought 6 file cabinets at $15 each. How much change did he receive from $100?
   - (1) add then divide
   - (2) multiply then subtract
   - (3) divide then multiply
   - (4) add then multiply
   - (5) divide then subtract

3. Socks were on sale at 5 pairs for $15. The clerk allowed Rudy to buy just 3 pairs at the sale rate. How much did Rudy spend on socks?
   - (1) subtract then multiply
   - (2) multiply then divide
   - (3) subtract then multiply
   - (4) divide then multiply
   - (5) add then multiply

4. Millen used 15 gallons of gasoline to drive 300 miles. She has another 160 miles to drive. How many more gallons of fuel will she use?

(1) divide then add
(2) divide then subtract
(3) divide then divide
(4) divide then multiply
(5) add then divide

5. Sarah had a piece of electrical wire that was 8 feet long. She had another piece that was 2 times as long as the first piece. How many feet of wire did she have altogether?

(1) divide then add
(2) multiply then divide
(3) multiply then subtract
(4) subtract then multiply
(5) multiply then add

6. The original price of a lamp was $35. During a sale, the price was reduced by $7. Glen bought 2 of the lamps. How much did he pay altogether?

(1) subtract then multiply
(2) multiply then subtract
(3) add then multiply
(4) divide then multiply
(5) divide then add

Part B. Choose the answer that gives the solution to each problem.

1. Louise works at a part-time job for 3 hours each day. She earns $5 per hour. How much money does she earn in 5 days?

(1) $15
(2) $25
(3) $40
(4) $ 75
(5) $200

2. Jonah bought 6 file cabinets at $15 each. How much change did he receive from $100?

(1) $10
(2) $25
(3) $85
(4) $90
(5) $94

3. Socks were on sale at 5 pairs for $15. The clerk allowed Rudy to buy just 3 pairs at the sale rate. How much did Rudy spend on socks?

(1) $ 5
(2) $ 9
(3) $15
(4) $27
(5) $45

4. Millen used 15 gallons of gasoline to drive 300 miles. She has another 160 miles to drive. How many more gallons of fuel will she use?

(1)   5 gallons
(2)   8 gallons
(3) 23 gallons
(4)   30 gallons
(5) 2400 gallons

5. Sarah had a piece of electrical wire that was 8 feet long. She had another piece that was 2 times as long as the first piece. How many feet of wire did she have altogether?

(1)   8 feet
(2) 10 feet
(3) 16 feet
(4) 24 feet
(5) 32 feet

6. The original price of a lamp was $35. During a sale, the price was reduced by $7. Glen bought 2 of the lamps. How much did he pay altogether?

   (1) $10
   (2) $14
   (3) $28
   (4) $56
   (5) $84

Check your answers on page 227.

WHOLE-NUMBERS REVIEW

Part A. Solve each problem.

1. $53 + 26 =$ 2. $648 - 314 =$ 3. $42 \times 3 =$

4. $81 \div 9 =$ 5. $5651 - 50 =$ 6. $84 - 65 =$

7. $20 \times 8 =$ 8. $84 + 11 =$ 9. $96 \div 3 =$

10. $4575 \div 5 =$ 11. $18 \times 6 =$ 12. $79 + 40 =$

13. $39 + 93 =$ 14. $84 - 65 =$ 15. $39 \times 59 =$

16. $2368 \div 32 =$ 17. $93 - 35 =$ 18. $7135 \times 92 =$

19. $1742 \div 66 =$ 20. $606 + 262 =$ 21. $42{,}285 - 1{,}991 =$

22. $4578 \div 327 =$ 23. $2405 \times 30 =$ 24. $268 + 593 =$

25. $6308 - 2409 =$ 26. $435 + 65 =$ 27. $40{,}041 - 8{,}252 =$

28. $6234 \times 503 =$ 29. $17{,}500 \div 100 =$ 30. $7128 + 76 + 402 =$

31. $20{,}805 \div 625 =$ 32. $30{,}000 \times 720 =$ 33. $874{,}644 - 229{,}381 =$

Part B. Solve each problem.

1. Write twenty-two thousand, two hundred twenty-one as a number.
2. Write 415,632 in words.
3. Round 374,959 to the nearest ten thousand.
4. Which digit is in the hundreds place in the number 32,476?
5. Round 2894 to the nearest thousand.
6. Find the sum of 312, 604, 515, and 11,003.
7. Find the quotient of 4815 divided by 15.
8. Find the average of these numbers: 38, 59, 67, 93, 76, 45.
9. Find the perimeter of a triangle whose sides are 12 inches, 12 inches, and 1 foot.
10. Find the difference between 16,030 and 8,873.
11. Find the product of 468 and 23.
12. Find the area of a rectangle whose length is 42 feet and whose width is 24 feet.

Check your answers on page 227.

This section will give you practice in answering questions like those on the GED. The Mathematics Test of the GED has 56 multiple-choice questions. Each question has five choices. The 14 questions in this Practice are all multiple-choice, like the ones on the GED. As you do this Practice, use the skills you've studied in this unit, especially the skills for solving whole-number word problems:

- Understand the question.
- Find the facts you need to solve the problem.
- Decide which math operation to use.
- Estimate the solution.
- Solve the problem and check your answer.

Directions: Choose the <u>one best answer</u> to each item.

1. The Cook Company employs 87 managers, 302 people in its manufacturing department, 234 people in sales, and 106 engineers. What is the total number of people employed by the Cook Company?

   (1) 619
   (2) 621
   (3) 720
   (4) 729
   (5) 61,119

2. Last year Ewing Township High School had 3894 students. This year the number of students increased by 766. How many students are enrolled this year?

   (1)   5 r64
   (2)   2,330
   (3)   3,818
   (4)   4,660
   (5) 2,982,804

3. Trailbreaker, a new four-wheel-drive car, lists at $19,034. Its chief competitor sells for $17,250. What is the difference in price between the two cars?

   (1) $ 1,784
   (2) $ 2,224
   (3) $ 2,884
   (4) $18,142
   (5) $36,284

4. Last Saturday 2532 people attended the opening of a new play. Before the play ended, 460 people had left. How many people remained for the entire play?

   (1)   5 r232
   (2) 2072
   (3) 2172
   (4) 2532
   (5) 2992

5. Maria drives the 1211 miles between Philadelphia and New Orleans quite often. In fact, she drove that distance 12 times in the past 6 months. How many miles did she travel in all?

   (1) 87,192
   (2) 14,532
   (3)   7,266
   (4)   3,633
   (5)   100

6. In a recent year a town spent $4640 to educate each student in its elementary schools. There were 809 students in the schools. What was the total education budget for elementary school students that year?

   (1) $73,161,760
   (2) $ 3,753,760
   (3) $ 78,880
   (4) $ 4,640
   (5) $ 3,831

7. Four people won a $500,000 lottery. They shared the prize equally. How much did each person receive?

   (1) $ 12,500
   (2) $ 50,000
   (3) $ 125,000
   (4) $ 500,000
   (5) $2,000,000

8. Tim Kau drove 216 miles on 18 gallons of gasoline in his truck. How many miles did Tim Kau drive on one gallon of gasoline?

   (1)   12
   (2) 117
   (3) 198
   (4) 234
   (5) 3888

9. Coach Jackson's starting basketball team included a center who was 6 feet 10 inches, forwards who were 6 feet 2 inches and 6 feet 1 inch, and guards who were 5 feet 9 inches and 5 feet 2 inches. What was the average height of the players?

   (1)   5 feet 5 inches
   (2)   6 feet
   (3)   6 feet 5 inches
   (4) 28 feet 5 inches
   (5) 30 feet

10. What is the perimeter of a square whose sides are each 1 foot 2 inches long?

   (1)   2 feet 4 inches
   (2)   3 feet 6 inches
   (3) 46 inches
   (4)   4 feet 8 inches
   (5) 16 feet 4 inches

11. In the past 5 years, attendance at the first football game of the season at State University was as follows: 40,963; 37,461; 40,066; 34,312; and 50,023. What was the average opening-day attendance for those three years?

   (1)   40,400
   (2)   40,565
   (3)   202,000
   (4)   202,825
   (5)   1,014,125

Item 12 is based on the following figure.

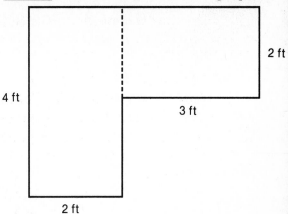

4 ft   2 ft   3 ft   2 ft

12. The figure shows the top of a desk. What is the total area of the desk top?

   (1)   6 square feet
   (2) 11 square feet
   (3) 14 square feet
   (4) 18 square feet
   (5) 48 square feet

13. Cary took $150 to shop. She bought a blouse for $24, a jacket for $49, and a skirt for $33. She put $8 worth of gasoline in her car and spent $4 for lunch. How much money did Cary have left?

   (1) $118
   (2) $ 52
   (3) $ 42
   (4) $ 36
   (5) $ 32

14. Four sacks of concrete weigh 160 pounds. How much do 18 sacks of concrete weigh?

   (1)   40 pounds
   (2)   72 pounds
   (3)   142 pounds
   (4)   720 pounds
   (5)   2880 pounds

Check your answers on page 228.

GED PRACTICE 1 SKILLS CHART

To review the mathematics skills covered by the items in GED Practice 1, study the following lessons in Unit 1.

Unit 1	Whole Numbers	Item Number
Lesson 6	Carrying in Addition	1, 2
Lesson 8	Borrowing in Subtraction	3, 4
Lesson 9	Multiplying Whole Numbers	5
Lesson 10	Shortcuts in Multiplication	6
Lesson 11	Dividing Whole Numbers	7
Lesson 12	Dividing by Two- and Three-Digit Numbers	8
Lesson 14	Working with Units of Measure: The English System	9, (10)
Lesson 15	Finding Perimeter and Area	10
Lesson 16	Working with Perimeter and Area	12
Lesson 17	Finding Averages	(9), 11
Lesson 19	Solving Multistep Whole-Number Word Problems	13, 14

UNIT 2

Decimals

Chapter 1 of this unit introduces decimals. In Chapters 2 and 3, you will practice addition, subtraction, multiplication, and division using decimals. In Chapter 4 you will use decimals to solve everyday problems— problems like many of those on the GED. You will work with word problems; metric measurements; averages; perimeter, circumference, and area; cost; and multistep problems.

Unit 2 Overview
Chapter 1   Understanding Decimals
Chapter 2   Addition and Subtraction
Chapter 3   Multiplication and Division
Chapter 4   Using Decimals

GED Practice 2

1 UNDERSTANDING DECIMALS

This chapter introduces decimals. It covers place value, reading and writing decimals, comparing and ordering decimals, and rounding decimals.

Lesson 20

Place Value

Decimals name values less than 1. Decimals break one thing down into 10 parts, or 100 parts, or 1000 parts, and so on.

Whether you realize it or not, you are already familiar with decimals. The system of money in the United States is based on decimals. When you spend $.89 for green peppers, or $15.37 for a turkey, you are dealing with decimals. Similarly, you might buy 4.2 gallons of gasoline for your car. The amount of gasoline, 4.2 gallons, has a decimal in it.

The period in each of these examples is the **decimal point**. Digits to the left of a decimal point are part of a whole number. Digits to the right of a decimal point are part of a decimal.

Whole Numbers   Decimals
.89
15.37
4.2
↑
Decimal points

Decimal Places

Every place or position in a decimal has a certain value. The chart that follows shows the first six decimal places.

Place Names	tenths	hundredths	thousandths	ten thousandths	hundred thousandths	millionths
Places .	___	___	___	___	___	___

Value gets smaller. →

As you move to the right from the decimal point, the value of the decimal places decreases. That is, each place is 10 times smaller than the one on its left. Think of it in terms of money. Consider the value of each digit in $.22.

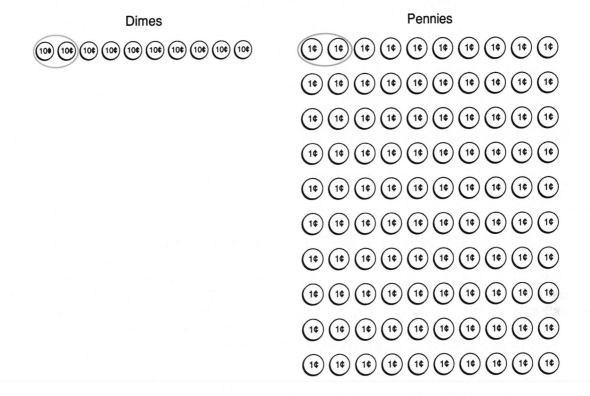

Dimes

Pennies

In $.22, the 2 nearest the decimal stands for 2 tenths of a dollar, or two dimes. The 2 in the second decimal place stands for 2 hundredths of a dollar, or two pennies.

To find the values of the digits in a decimal, find which decimal place each digit is in.

Example 1: What are the values of the digits in .972?

tenths
hundredths
thousandths
.972

The 9 is in the tenths place, so it has a value of 9 tenths.
The 7 is in the hundredths place, so it has a value of 7 hundredths.
The 2 is in the thousandths place, so it has a value of 2 thousandths.

Part A. Answer the following questions.

1. Which digit is in the tenths place in each number?
   (a) .51   (b) .125   (c) .457   (d) .2379   (e) .75

2. Which digit is in the hundredths place in each number?
   (a) .837   (b) .73   (c) .3784   (d) .783   (e) .842

3. Which digit is in the thousandths place in each number?
   (a) .3492   (b) .99432   (c) .344229   (d) .92939   (e) .42234

Part B. Complete each of the following statements.

1. .52 has ___ tenths, and ___ hundredths.

2. .265 has ___ tenths, ___ hundredths, and ___ thousandths.

3. .8925 has ___ tenths, ___ hundredths, ___ thousandths, and ___ ten thousandths.

4. .98523 has ___ tenths, ___ hundredths, ___ thousandths, ___ ten thousandths, and ___ hundred thousandths.

5. .673127 has ___ tenths, ___ hundredths, ___ thousandths, ___ ten thousandth, ___ hundred thousandths, and ___ millionths.

Check your answers on page 228.

Zeros in Decimals

Just as in whole numbers, zeros in decimals have no value. They hold places.

Example 2: What are the values of the digits in .0201?

```
          tenths
          hundredths
          thousandths
          ten thousandths
      .0 2 0 1
```

Neither of the 0's has any value at all.
The 2 is in the hundredths place, so it has a value of 2 hundredths.
The 1 is in the ten-thousandths place, so it has a value of 1 ten thousandth.

Complete each of the following statements.

1. In .08, the 8 is in the _____ place.

2. In .081, the 0 holds the _____ place.

3. In .002003, the 2 is in the _____ place, and the 3 is in

   the _____ place.

4. .017 has ___tenths, ___ hundredths, and ___ thousandths.

5. In .09, the digit in the tenths place is a ___.

Check your answers on page 229.

Reading and Writing Decimals

Reading Decimals

To read a decimal, say the number as if it were a whole number. Then say the name of the last place in the decimal.

Example 1: Read .836.

STEP 1: Say the decimal as if it were a whole number: "eight hundred thirty-six."

STEP 2: Say the name of the last place in the decimal: "thousandths."

The decimal is read *eight hundred thirty-sixth thousandths.*

When a decimal begins with one or more zeros, ignore them when you say the number. For example, .067 is read *sixty-seven thousandths.*

Read these decimals.

1. .42	2. .35	3. .82	4. .71	5. .98
6. .4	7. .613	8. .03	9. .763	10. .003
11. .2112	12. .2104	13. .909	14. .2571	15. .0007

16. .3245   17. .12   18. .412   19. .078   20. .53208

21. .4   22. .22364   23. .000014   24. .3167   25. .000006

Check your answers on page 229.

When there is a whole number to the left of the decimal point, as in 25.65, it is called a **mixed decimal**. When reading mixed decimals, the decimal point is read as the word *and*.

Example 2: Read 4.123.

STEP 1: Say the whole number: "four."

STEP 2: Read the decimal point: "and."

STEP 3: Say the decimal part of the number: "one hundred twenty-three thousandths."

The mixed decimal is read *four and one hundred twenty-three thousandths*.

EXERCISE 21b

Part A. Match each number in column A with the words in column B.

Column A	Column B
____ 1. 1.09	(a) fifty-nine and nine thousand five hundred nine ten thousandths
____ 2. 6.5	(b) four and twenty-three thousandths
____ 3. 4.023	(c) one and nine hundredths
____ 4. 3000.303	(d) six and five tenths
____ 5. 59.9509	(e) three thousand and three hundred three thousandths
____ 6. 2.05	(f) eighteen and four tenths
____ 7. 18.4	(g) ninety-nine and ninety-nine hundredths
____ 8. 73.73	(h) two and five hundredths
____ 9. 99.99	(i) seventy-three and seventy-three hundredths
____ 10. 9.99	(j) nine and ninety-nine hundredths

Part B. Write the following numbers in words.

1. .36 _____

2. .9 _____

3. .04 _____

4. .52 _____

5. .37 _____

6. .478 _____

7. .0021 _____

8. 16.478 _____

9. 2.6 _____

10. 21.04 _____

WORD PROBLEM

A machinist had to cut a small piece from a copper pipe. The piece had to be exactly .15 inch long. Express the length of the piece in words.

Check your answers on page 229.

Writing Decimals

To write a decimal in digits, you need to decide how many decimal places there are in the number. That way, you'll know if you need to write any zeros next to the decimal point.

Example 3: Write nine and forty-eight thousandths in digits.

Step 1	Step 2	Step 3
48	.048	9.048

STEP 1: Write the 48.

STEP 2: Because the decimal has the word *thousandths* in it, the decimal needs three places. Write a zero before the 48. Write the decimal point before the zero.

STEP 3: Write the whole number, 9.

Write each of the following in digits.

1. six hundredths _____
2. seven thousandths _____
3. eight ten thousandths _____
4. forty-eight hundred thousandths _____
5. thirty-six thousandths _____
6. one and nine thousandths _____
7. two and thirty-six ten thousandths _____
8. twenty and two hundredths _____
9. eight hundred three and eighty-three thousandths _____
10. seventy-seven and seventy-seven hundredths _____
11. three thousandths _____
12. thirty-eight and eight hundred three ten thousandths _____
13. one hundred fifty and fifteen thousandths _____
14. eighty and eight hundredths _____
15. six thousand fifty and fifty-six thousandths _____

WORD PROBLEM

Rudy asked Dave to put exactly three and three tenths gallons of gas in a container. If the gas pump used a mixed decimal to show the amount of gas pumped, what did the pump show after Dave gave Rudy the amount of gas he asked for?

Check your answers on page 229.

Lesson
22

Comparing and Ordering Decimals

Comparing Decimals

When you have two or more decimals, it is not always easy to see which is the largest and which is the smallest. A method that makes comparing decimals easier is to write them in a column with the decimal points lined up. Then, starting with the digits in the tenths place of each decimal, compare the sizes of the digits.

Example 1: Which is larger, .523 or .524?

Step 1	Step 2	Step 3	Step 4
.523	.523	.523	.523
.524	.524	.524	.524

STEP 1: Write the decimals in a column, lining up the decimal points.

STEP 2: Compare the digits in the tenths places of both decimals. They are the same. Both are 5.

STEP 3: Compare the digits in the hundredths places. They are also the same. Both are 2.

STEP 4: Compare the digits in the thousandths places. 4 is larger than 3. Therefore, .524 is larger than .523.

When two decimals you are comparing have a different number of places, it helps to write one or more zeros at the end of the decimal with the fewest places. The zeros have no value, but both decimals then have the same number of places.

Example 2: Which is smaller, .13 or .132?

Step 1	Step 2	Step 3	Step 4
.130	.130	.130	.130
.132	.132	.132	.132

STEP 1: Write the decimals in a column, lining up the decimal points. Write a 0 at the end of .13 so that it has as many places as .132.

STEP 2: Compare the digits in the tenths places of both decimals. They are the same. Both are 1.

STEP 3: Compare the digits in the hundredths places. They are also the same. Both are 3.

STEP 4: Compare the digits in the thousandths places. 0 is smaller than 2. Therefore, .13 is smaller than .132.

EXERCISE 22a

Part A. Find the larger decimal in each pair.

1. .6   .066

2. .505 .55

3. .011   .1

4. .75 .76	5. .90 .099	6. .320 .3
7. .582 .73	8. .48 .048	9. .04 .15
10. .088 .18	11. .40 .401	12. .321 .6
13. 5.25 5.02	14. 17.5 17.3	15. 89.09 89.8
16. 4.07 4.5		

Part B. Find the smaller decimal in each pair.

1. .04 .044	2. .2 .12	3. .3 .321
4. 0.60 0.59	5. .34 .312	6. .489 .498
7. .6 .5	8. .8 .1	9. .232 .125
10. .22 .234	11. .34 .312	12. .07 .09
13. 1.066 1.093	14. 4.05 4.5	15. 3.04 3.11
16. 21.01 20.7		

WORD PROBLEM

Fred used a micrometer to measure the thickness of two different sheets of paper. One was .025 inch thick. The other was .009 inch thick. Which sheet was thicker?

Check your answers on page 229.

Ordering Decimals

You may need to write several decimals or mixed decimals in order from largest to smallest or from smallest to largest. The following example shows a quick way to do that.

Example 3: Write 7.32, 7.231, and 7.312 in order from the smallest to the largest.

Step 1	Step 2	Step 3	Step 4
7.320	7.320	7.320	7.320
7.231	7.231	7.231	
7.312	7.312	7.312	7.312

STEP 1: Write the mixed decimals in a column, lining up the decimal points. Write a 0 at the end of 7.32 so that it has three places like the other two numbers.

STEP 2: Compare the digits in the ones places. They are all the same: 7.

STEP 3: Compare the digits in the tenths places. One of them is a 2, and the others are both 3. The mixed decimal with the 2 in the tenths place is, therefore, the smallest number.

STEP 4: Compare the digits in the hundredths places of the two remaining numbers. One of them is a 1, and the other is a 2. The mixed decimal with the 1 in the hundredths place is, therefore, the next smallest number. In this case, it is not necessary to compare the digits in the thousandths places because you already know that, in order from smallest to largest, the numbers are 7.231, 7.312, and 7.32.

EXERCISE 22b

Part A. Put each set of decimals and mixed decimals in order from largest to smallest.

1. 3.28   4.01   3.34   2. 7.41   7.34   8.4

3. 5.91   5.87   5.9   4. 2.69   3.96   2.8

5. 5.73   5.37   7.35   6. 2.3   2.15   2.55

7. 3.088   3.08   3.07   8. .693   .6   .655

Part B. Put each set of decimals and mixed decimals in order from smallest to largest.

1. 1.7   2.5   3.9   2.7   2. 2.43   4.13   4.23   3.4

3. .6   .821   1.68   1.02   4. 43.7   34.7   40.3   34.07

5. 9.09   9.9   .99   9.99   6. 4.89   4.983   4.980   4.889

7. .64   .60   .604   .641   8. .28   .23   .229   .234

WORD PROBLEM

Terry measured the amount of rainfall on 4 days in April. On the first day, 3.12 inches of rain fell. On the second day, .87 inch fell. The third day there were 3.02 inches, and the fourth day there were 3.14 inches. On which day did the most rain fall? On which day did the least rain fall?

Check your answers on page 230.

Rounding Decimals

Many times it is useful to estimate, or round decimals and mixed decimals. For example, you may want to roast a 6.45-pound chicken for about 20 minutes per pound. To figure out how long to roast the chicken, you might estimate its weight at 6.5 pounds. If you do, you are rounding the weight to the nearest tenth.

Rounding decimals is similar to rounding whole numbers. You will probably remember the first two of these steps. The last step is used only with decimals.

- Underline the digit in the place you want to round to.
- Is the digit to the right of the one you underlined less than 5? If it is, leave the underlined digit as it is. Is the digit to the right of the one you underlined 5 or more than 5? If it is, add 1 to the underlined digit.
- Drop the digits to the right of the underlined digit.

Example 1: Round .1392 to the nearest hundredth.

Step 1	Step 2	Step 3
	4	
.1392	.1392	.14

STEP 1: Underline the digit in the hundredths place—the 3.

STEP 2: Look at the digit to the right of the underlined 3. Since 9 is more than 5, change the underlined 3 to a 4.

STEP 3: Drop the digits to the right of the 4. Rounded to the nearest hundredth, .1392 is .14

EXERCISE 23a

Part A. Round each of the following decimals and mixed decimals to the nearest tenth.

1. 4.46

2. 15.31

3. 48.08

4. 224.55

5. 96.324

6. .294

7. .355

8. .4321

9. .77890

10. .44

Part B. Round each of the following decimals and mixed decimals to the nearest hundredth.

1. .6378

2. .6738

3. .845

4. .397

5. .4432

6. 4.264

7. 9.137

8. 14.063

9. 68.345

10. 156.051

Part C. Round each of the following decimals and mixed decimals to the nearest thousandth.

1. .2696

2. .6778

3. 2.43012

4. 9.89467

5. 82.6007

6. 1.8855

7. 23.2235

8. 8.3124

9. 76.0047

10. 11.2345

WORD PROBLEM

When Jason placed a fish he'd caught on a digital scale, the weight registered 7.45 pounds. When he told his friends how much the fish weighed, he rounded the weight to the nearest tenth of a pound. What did he tell his friends the fish weighed?

Check your answers on page 230.

Sometimes decimals and mixed decimals are rounded to the nearest whole number. That is the same as rounding to the ones place.

Example 2: Round 7.56 to the nearest whole number.

Step 1	Step 2	Step 3
	8	
7.56	7.56	8

STEP 1: Underline the digit in the ones place—the 7.

STEP 2: Look at the digit to the right of the underlined 7. Since it is 5, change the underlined 7 to an 8.

STEP 3: Drop the decimal point and the digits to the right of the 8. Rounded to the nearest whole number, 7.56 is 8.

Round each of the following amounts to the nearest whole number.

1. 5.8   2. 15.14   3. 49.365   4. 12.9

5. 6.5   6. 12.44   7. 4.568   8. 32.009

9. 16.789   10. 2.475   11. $6.27   12. $3.55

13. $76.99   14. $11.48   15. $3.64   16. 8.0967

17. 23.54   18. 1265.09   19. 14.458   20. 56.98

WORD PROBLEM

Jeff has exactly $5.89 in his pocket. How much is that rounded to the nearest dollar?

Check your answers on page 230.

Chapter

2 ADDITION AND SUBTRACTION

This chapter covers addition and subtraction of decimals and mixed decimals. Lesson 24 covers addition skills, Lesson 25 covers subtraction skills.

Lesson 24

Adding Decimals and Mixed Decimals

To add decimals and/or mixed decimals, write them in a column with the decimal points lined up. Then add as you would whole numbers. Finally, place a decimal point in the answer under the decimal points in the problem.

When decimals do not have the same number of places, you may find it helpful to write zeros as placeholders. Writing zeros to the end of a decimal does not change the decimal's value. The zeros simply help you to keep the columns straight.

Example 1: Find the sum of 1.3, .04, and .425.

Step 1	**Step 2**	**Step 3**
1.300	1.300	1.300
.040	.040	.040
+ .425	+ .425	+ .425
	1 7 6 5	1.765

STEP 1: Set up the problem in a column with the decimal points lined up. (If you like, write zeros to the right of .3 and .04 so that all the decimals have the same number of places.

STEP 2: Add.

STEP 3: Place the decimal point in the answer under the decimal points in the problem.

Add the following decimals and mixed decimals.

1. 2.4 + 1.1 =   2. .3 + .6 =   3. $.35 + $.43 =

4. 7.77 + 1.11 =   5. $.82 + $.17 =   6. .1 + 2.008 =

7. .124 + 3.04 =   8. .54 + .32 =   9. $.31 + $.18 =

10. 6.41 + .122 =   11. 8.2 + 1.11 + .003 =   12. .66 + 9.3 =

13. .3 + .12 =   14. .011 + .33 + .35 =   15. .32 + .4 + 5.11 =

WORD PROBLEM

The key word *altogether* tells you to add to solve the problem.

Mrs. Ho bought two small steaks. One steak weighed .37 pound. The other steak weighed .42 pound. How much did the steaks weigh **altogether**?

Check your answers on page 230.

Carrying

Carry when you add decimals or mixed decimals just as you would when you add whole numbers. When you carry from the tenths place to the ones place, ignore the decimal points in the problem.

Example 2:   Find the sum of 1.61 and .79.

Step 1	**Step 2**	**Step 3**

STEP 1: Set up the problem in a column with the decimal points lined up.

STEP 2: Add, carrying as necessary.

STEP 3: Place the decimal point in the answer under the decimal points in the problem. The answer is 2.40, or 2.4.

When a decimal ends with a zero, you do not need to write the zero because it has no value. That is why the answer to the problem in Example 2 can be either 2.40 or 2.4.

EXERCISE 24b

Add the following decimals and mixed decimals.

1. .4 + .8 =

2. .6 + .7 =

3. .35 + .56 =

4. $.75 + $.96

5. .12 + .79 =

6. .82 + .2 =

7. .76 + .35 =

8. .241 + .139 =

9. .42 + .21 =

10. .606 + .306 =

11. .44 + .25 =

12. .75 + .08 =

13. .32 + .21 + .39 =

14. .11 + .74 + .42 =

15. .09 + .14 + .52 =

16. 3.28 + 1.35 =

17. 86.36 + 5.47 =

18. 118.18 + 18.18 =

19. 2.93 + 3.42 =

20. 6.82 + 5.44 =

21. 10.24 + 6.94 =

22. 2.658 + 1.848 =

23. 4.857 + 3.274 =

24. 3.019 + 27.179 =

25. $28.69 + $14.75 =

26. $2.77 + $6.20 =

27. 7.8 + 10.01 =

28. .555 + .476 =

29. .498 + .312 =

30. .384 + .108 =

31. .741 + .059 =

32. .333 + .676 =

33. .845 + .1367 =

34. .6008 + .9123 =

35. .4545 + .2887 =

36. .7471 + .2829 =

37. 126.003 + 512.6 + 19.7 =

38. 1.20 + 5.1 + 7.34 =

39. 900.001 + 76.8 + 83.475 =

40. .637 + .093 =

WORD PROBLEM

The key words *in all* tell you to add to solve the following problem.

Salinda spent $.70 for coffee on Monday. On Tuesday, she spent $1.20 for coffee and a roll. On Wednesday, she bought a soda for $.85. How much did she spend **in all** those 3 days?

Check your answers on page 230.

Adding Whole Numbers to Decimals and Mixed Decimals

When adding a decimal or a mixed decimal to a whole number, write a decimal point to the left of the whole number. Then write zeros as necessary to hold place value.

Example 3: Find the sum of $18, $.64, and $1.21.

Step 1	**Step 2**	**Step 3**
tens ones tenths hundredths	tens ones tenths hundredths	ones tenths tenths hundredths
$18.00	$18.00	$18.00
.64	.64	.64
+ 1.21	+ 1.21	+ 1.21
	$19 85	$19.85

STEP 1: Set up the problem in a column with the decimal points lined up. Write a decimal point and two zeros to the right of $18, the whole number.

STEP 2: Add.

STEP 3: Place the decimal point in the answer under the decimal points in the problem.

EXERCISE 24c

Add the following decimals and mixed decimals and whole numbers.

1. $45 + $3.23 =

2. 10 + .072 =

3. 100 + .05 =

4. 303 + 5.03 =

5. .54 + 5.4 + 54 =

6. 96 + 34 + .01 =

7. 42 + 3.45 + .67 =

8. 25.5 + .36 + 44 =

9. 396 + .003 =

10. 774 + .631 + 99 =

11. 55.55 + .55 + 42 =

12. 12 + 98 + .07 =

13. 67 + .348 + 21 =

14. 11 + 88 + .77 =

15. 43 + .34 + .56 =

Check your answers on page 230.

Lesson 25

Subtracting Decimals and Mixed Decimals

To subtract decimals and/or mixed decimals, write them in a column with the decimal points lined up. When decimals do not have the same number of places, write zeros as placeholders. (Remember: Writing zeros at the end of a decimal does not change the decimal's value, but it helps you keep the columns straight.) Then subtract as you would whole numbers. Finally, place a decimal point in the answer under the decimal points in the problem.

Example 1: Find the difference between 22.431 and 10.2.

Step 1	Step 2	Step 3
tens ones tenths hundredths thousandths	tens ones tenths hundredths thousandths	tens ones tenths hundredths thousandths
2 2 . 4 3 1	2 2 . 4 3 1	2 2 . 4 3 1
− 1 0 . 2 0 0	− 1 0 . 2 0 0	− 1 0 . 2 0 0
	1 2 2 3 1	1 2 . 2 3 1

STEP 1: Set up the problem in a column with the decimal points lined up. Be sure to put the first number on top. Write zeros to the right of the 2 in 10.2 so that both numbers have the same number of decimal places.

STEP 2: Subtract.

STEP 3: Place the decimal point in the answer under the decimal points in the problem.

Subtract to solve the following problems.

1. .7 − .3 =

2. .8 − .5 =

3. $.44 − $.13 =

4. .75 − .22 =

5. $.89 − $.37 =

6. .45 − .34 =

7. $.95 − $.73 =

8. .742 − .341 =

9. .509 − .306 =

10. .895 − .573 =

11. $.89 − $.51 =

12. .37 − .2 =

13. .365 − .12 =

14. .0845 − .034 =

15. .9274 − .705 =

16. .57 − .23 =

17. .279 − .22 =

18. .0048 − .003 =

19. .993 − .81 =

20. .666 − .44 =

21. $68.93 − $4.91 =

22. 74.989 − 1.773 =

23. 14.6 − 3.2 =

24. $29.87 − $.75 =

25. 5.895 − .063 =

26. $35.77 − $.56 =

27. 67.5521 − 23.4511 =

28. 3.719 − .719 =

29. 7.503 − 6.403 =

30. 6.7703 − 6.6702 =

31. 27.75 − 6.32 =

32. 12.07 − 1.04 =

33. $89.74 − $78.63 =

34. 64.32 − 30.21 =

35. 18.8 − 6.4 =

36. 112.004 − 1.002 =

37. .54 − .32 =

38. 82.28 − 60.04 =

39. 900.08 − 700.01 =

40. 36.468 − 23.246 =

WORD PROBLEM

The key words *decreased . . . by* tell you to subtract to solve the following problem.

The regular price of the sweaters at the Ruffled Rag is $39.95. During a sale, the store **decreased** the price **by** $6.50. What was the sale price of the sweaters?

Check your answers on page 230.

Borrowing

Borrow when you subtract decimals or mixed decimals just as you would when you subtract whole numbers. When you borrow from the ones place for the tenths place, ignore the decimal points in the problem.

Example 2: Find the difference between 1.61 and .79.

<table>
<tr><th>Step 1</th><th>Step 2</th><th>Step 3</th></tr>
<tr>
<td>

```
    ones
    tenths
    hundredths
    1 . 6 1
  -   . 7 9
```

</td>
<td>

```
    ones
    tenths
    hundredths
         15 11
    1̸ . 6̸ 1̸
  -   . 7 9
         8  2
```

</td>
<td>

```
    ones
    tenths
    hundredths
         15 11
    1̸ . 6̸ 1̸
  -   . 7 9
      .  8  2
```

</td>
</tr>
</table>

STEP 1: Set up the problem in a column with the decimal points lined up.

STEP 2: Subtract, borrowing as necessary.

STEP 3: Place the decimal point in the answer under the decimal points in the problem.

EXERCISE 25b

Subtract to solve the following problems. Borrow as necessary.

1. 6.2 − 3.3 =

2. 20.6 − 10.7 =

3. 312.6 − 1.7 =

4. $7.54 − $2.98 =

5. 6.2 − 4.25 =

6. 38.06 − 15.87 =

7. $87.98 − $49.99 =

8. 40.004 − 22.407 =

9. 9.86 − 1.189 =

10. 82.008 − 7.021 =

11. 7.8762 − 3.2958 =

12. 372.2 − 28.14 =

13. 4.1206 − 2.517 =

14. 220.8 − 13.13 =

15. 8.678 − 4.99 =

16. 46.61 − 17.24 =

17. 38.74 − 9.85 =

18. 16.5 − 9.6 =

19. $15.36 − $13.48 =

20. $42.42 − $24.42 =

WORD PROBLEM

The key word *less* tells you to subtract to solve the following problem.

The weight of some electronic parts is measured in thousandths of an ounce. Mr. Yee's calculator has two electronic chips. One weighs .243 ounce, and the other weighs .05 ounce. How much **less** does the lighter chip weigh?

Check your answers on page 231.

Decimal Subtraction with Whole Numbers

In a subtraction problem that contains a decimal or a mixed decimal and a whole number, write a decimal point and zeros to the left of the whole number. The zeros hold place value.

Example 3: Find the difference between 43 and .25.

Step 1	**Step 2**	**Step 3**

```
       tens              tens              tens
       ones              ones              ones
       tenths            tenths            tenths
       hundredths        hundredths        hundredths

                              9                 9
                          2  10 10          2  10 10
    4 3 . 0 0         4 3 . 0 0         4 3 . 0 0
  -     . 2 5       -     . 2 5       -     . 2 5
                      4 2   7 5         4 2 . 7 5
```

STEP 1: Write a decimal point and two zeros to the right of 43, the whole number. Set up the problem in a column with the decimal points lined up.

STEP 2: Subtract, borrowing as necessary.

STEP 3: Place the decimal point in the answer under the decimal points in the problem.

EXERCISE 25c

Subtract to find the solutions to the following problems.

1. $84 - .62 =$ 2. $75 - .05 =$ 3. $68 - .68 =$

4. $23 - .13 =$ 5. $11 - .03 =$ 6. $96 - .89 =$

7. $42 - .31 =$ 8. $64 - .46 =$ 9. $88 - .067 =$

10. $36 - .554 =$ 11. $132 - .003 =$ 12. $312 - .042 =$

13. $656 - .778 =$ 14. $111 - .5 =$ 15. $231 - .089 =$

16. $623 - .623 =$ 17. $475 - .07 =$ 18. $8 - .3056 =$

19. $12 - .3104 =$ 20. $344 - .433 =$

WORD PROBLEM

The key word *difference* tells you to subtract to solve the following problem.

A saw usually costs $42. It is on sale now for $29.95. What is the **difference** between the regular price and the sale price?

Check your answers on page 231.

MIXED PRACTICE 3
ADDITION AND SUBTRACTION OF DECIMALS

These problems let you practice the decimal addition and subtraction skills you have learned. Read each problem carefully and solve it.

1. .537 + .22 + .041 =   2. .46 − .12 =   3. .478 − .25 =

4. 15 + .53 + 2.13 =   5. 4.3 + 3.2 =   6. 14.574 − 10.42 =

7. 19.42 − 7.59 =   8. .72 + .09 =   9. .737 − .402 =

10. 32 − .32 =   11. 24.6 + 32.8 =   12. 17.25 + 24.16 =

13. $5.05 − $.99 =   14. 16.2 + 3.4 =   15. 7.6 − 1.9 =

16. $9.49 − $6.32 =   17. 14.5 + 31.3 =   18. 3.789 + 2.009 =

19. 5.609 + .735 =   20. $.08 + $1.53 =   21. 8 − 7.52 =

22. 35.065 + 17.807 =   23. .53 − .16 =   24. 9.11 − 8.683 =

25. 8.92 − 1.96 =   26. 27.35 − 22.15 =   27. 7.01 − 3.5 =

28. 185 − 6.94 =   29. 160 − .052 =   30. 13 − 6.81 =

31. $7.03 − $4.01 =   32. $8.75 + $8.75 =   33. $6.09 + $5.22 =

34. 3.07 − .08 =   35. 9.00 − .75 =   36. .62 − .48 =

37. .503 − .39 =   38. 14.9 − .09 =   39. 28.506 − .5 =

40. $53.42 + $.57 + $5.36 =   41. 4.08 + 3.93 + 1.34 =

42. 33.308 + 59.874 =   43. $.10 + $.09 + $1.26 =

44. $3.00 + $7.98 + $6.34 =   45. 44.13 + 11.84 + 6.15 =

46. 45.312 + 31.666 =   47. 45.312 − 31.666 =

48. 20.342 − 16.605 =   49. $7.13 + $.89 + $12 =

50. $7.98 + $3 + $11 =

Check your answers on page 231.

Chapter

3 MULTIPLICATION AND DIVISION

This chapter covers multiplication and division of decimals and mixed decimals. Lesson 26 covers multiplication skills; Lesson 27 covers division skills.

Lesson 26

Multiplying Decimals and Mixed Decimals

You multiply decimals just as you multiply whole numbers. But you must place the decimal point correctly in the answer.

Multiplying a Decimal or a Mixed Decimal by a Whole Number

When you multiply a decimal or a mixed decimal by a whole number, the number of decimal places in the answer is the same as the number of places in the decimal or mixed decimal.

Example 1: Multiply .61 by 4.

Step 1	**Step 2**	**Step 3**
.61	.61	.61 2 decimal places
× 4	× 4	× 4
	244	2.44 2 decimal places

STEP 1: Set up the problem. (It is not necessary to line up decimal points when you multiply.)

STEP 2: Multiply.

STEP 3: There are two decimal places in .61, so there must be two decimal places in the answer. Place the decimal point two places from the right end of the answer: 2.44.

EXERCISE 26a

Part A. Place the decimal point correctly in the answer to each of the following multiplication problems, which have already been worked.

1. .50	2. .72	3. .11	4. 3.24	5. 16.211
× 2	× 4	× 3	× 6	× 3
100	288	33	1944	48633

6. .33	7. 2.473	8. .11	9. .14	10. .18
× 11	× 2	× 43	× 22	× 11
363	4946	473	308	198

Part B. Multiply to find the products.

1. $8.126 \times 4 =$ 2. $\$.34 \times 2 =$ 3. $\$1.04 \times 2 =$

4. $.9114 \times 8 =$ 5. $9.003 \times 3 =$ 6. $6.22 \times 3 =$

7. $\$.08 \times 9 =$ 8. $\$2.05 \times 1 =$ 9. $.213 \times 4 =$

10. $4.042 \times 2 =$ 11. $\$.34 \times 5 =$ 12. $.418 \times 5 =$

13. $.6753 \times 6 =$ 14. $\$1.04 \times 8 =$ 15. $6.8 \times 22 =$

16. $2.125 \times 5 =$ 17. $\$15.94 \times 2 =$ 18. $3.875 \times 3 =$

19. $.3675 \times 7 =$ 20. $1.814 \times 4 =$ 21. $.26 \times 4 =$

22. $3.61 \times 13 =$ 23. $.428 \times 17 =$ 24. $.81 \times 31 =$

25. $.64 \times 5 =$ 26. $\$2.65 \times 5 =$ 27. $.387 \times 43 =$

28. $.8768 \times 62 =$ 29. $.47 \times 7 =$ 30. $\$4.56 \times 12 =$

WORD PROBLEM

The key words *all four* tell you to multiply to solve the following problem.

Tomaso has 4 packages of ground beef. Each weighs .89 pound. What is the weight of **all four** packages?

Check your answers on page 231.

Multiplying a Decimal by a Decimal

When you multiply a decimal by a decimal, the number of decimal places in the answer is the same as the total number of places in the decimals multiplied.

Before you work a problem, estimating the answer can help you be sure that the answer you get when you work the problem makes sense. Your estimate and the answer should be close to each other in value.

Example 2:   Multiply .961 by .24.

Step 1	Step 2	Step 3	
.961	.961	.961	3 decimal places
× .24	× .24	× .24	2 decimal places
	3844	3844	
Estimate:	1922	1922	
1 × .25 = .25	23064	.23064	5 decimal places

STEP 1: Set up the problem. (Remember: It is not necessary to line up decimal points when you multiply.) You can estimate the answer by multiplying 1 × .25 = .25.

STEP 2: Multiply.

STEP 3: There are three decimal places in .961 and two decimal places in .24. Therefore, there must be five decimal places in the answer. Place the decimal point five places from the right end of the answer: .23064. Compare the answer to your estimate in Step 1: they are close in value.

When you multiply an amount of money by a decimal or a mixed decimal, it is often necessary to round the product to the nearest hundredth—that is, to the nearest cent. For example, if the product is $4.675, it should be rounded to $4.68.

EXERCISE 26b

Part A. Place the decimal point correctly in the answer to each of the following multiplication problems, which have already been worked.

1.	.632	2.	.6	3.	.567	4.	.91	5.	.4197
	× .64		× .3		× .25		× .9		× .68
	40448		18		14175		819		285396

Part B. Multiply to find the products. In problems that involve money, round answers to the nearest hundredth (the nearest cent) if the product has more than two decimal places.

1. .424 × .26 =   2. .42 × .4 =   3. .279 × .36 =

4. .65 × .7 =   5. $.43 × .5 =   6. .23 × .7 =

7. .9256 × .79 =   8. .7 × $.93 =   9. .44 × .6 =

10. .25 × $.50 =   11. .9752 × .8 =   12. .3646 × .413 =

Check your answers on page 231.

Multiplying a Mixed Decimal by a Decimal or a Mixed Decimal

When you multiply a mixed decimal by a decimal or a mixed decimal, the same rule applies: the number of decimal places in the answer is the same as the total number of places in the numbers multiplied.

Example 3: Multiply 9.61 by 2.4.

Step 1	**Step 2**	**Step 3**
9.61	9.61	9.61 2 decimal places
× 2.4	× 2.4	× 2.4 1 decimal place
	3844	3844
	1922	1922
	23064	23.064 3 decimal places

STEP 1: Set up the problem. (Remember: It is not necessary to line up decimal points when you multiply.)

STEP 2: Multiply.

STEP 3: There are two decimal places in 9.61 and one decimal place in 2.4. Therefore, there must be three decimal places in the answer. Place the decimal point three places from the right end of the answer: 23.064.

EXERCISE 26c

Part A. Place the decimal point correctly in the answer to each of the following multiplication problems, which have already been worked.

1. 6.32	2. 1.6	3. 56.7	4. 9.1	5. 41.97
× .64	× .3	× 2.5	× .9	× 6.8
40448	48	14175	819	285396

Part B. Multiply to find the products. Remember to round money answers to the nearest hundredth (the nearest cent).

1. $1.24 \times .12 =$
2. $\$1.31 \times .3 =$
3. $4.02 \times .4 =$

4. $26.1 \times .99 =$
5. $99.44 \times 1.2 =$
6. $\$.72 \times 1.1 =$

7. $2.2 \times 3.3 =$
8. $17.4 \times .321 =$
9. $.007 \times \$7.20 =$

10. $9.9 \times .83 =$

WORD PROBLEM

When a problem gives the rate of pay for one hour and asks you to find the pay for several hours, you can multiply to solve the problem.

As a clerk typist, Nina is paid $6.43 per hour. She works 37.5 hours per week. How much does she make a week?

Check your answers on page 232.

Zeros as Placeholders in Products

Sometimes it is necessary to write one or more zeros in front of a product so that there will be enough decimal places in the answer.

Example 4: Multiply .161 by .24.

Step 1	Step 2	Step 3	
.161	.161	.161	3 decimal places
× .24	× .24	× .24	2 decimal places
	644	644	
	322	322	
	3864	.03864	5 decimal places

STEP 1: Set up the problem. (Remember: it is not necessary to line up decimal points when you multiply.)

STEP 2: Multiply.

STEP 3: There are three decimal places in .161 and two decimal places in .24. Therefore, there must be five decimal places in the answer. Because there are only four digits in the product, you must write a zero in front of the product. Then you can place the decimal point five places from the right end of the answer: .03864.

Part A. Place the decimal point correctly in the answer to each of the following multiplication problems, which have already been worked.

1. .18	2. .1324	3. .789	4. .77	5. .19
× .23	× .55	× .003	× .11	× .11
414	72820	2367	847	209

Part B. Multiply to find the products. Remember to round money answers to the nearest hundredth (the nearest cent).

1. $.2 \times .3 =$

2. $\$.17 \times .5 =$

3. $.22 \times .03 =$

4. $.154 \times .06 =$

5. $.127 \times .65 =$

6. $.795 \times .002 =$

7. $.086 \times .04 =$

8. $\$.54 \times .005 =$

9. $.125 \times .032 =$

10. $.0069 \times .003 =$

11. $\$.83 \times .09 =$

12. $.389 \times .04 =$

13. $.19 \times .09 =$

14. $.616 \times .003 =$

15. $.823 \times .004 =$

16. $\$.08 \times .02 =$

WORD PROBLEM

When a problem gives the cost of one pound of something and asks you to find the cost of part of a pound, you can multiply to solve the problem.

To make soup, Lavatria bought .2 pound of bones that cost $.35 per pound. How much did she pay for the bones.

Check your answers on page 232.

Multiplying Decimals and Mixed Decimals by 10, 100, and 1000

To multiply a decimal or mixed decimal by 10, 100, or 1000, just move the decimal point one place to the right for every zero in the number you are multiplying by.

Example 5: Multiply 2.1234 by 10, by 100, and by 1000.

$$2.1234 \times \quad 10 = 21.234$$
$$2.1234 \times \quad 100 = 212.34$$
$$2.1234 \times 1000 = 2123.4$$

Sometimes it is necessary to write zeros at the right end of a number in order to move the decimal enough places. For example, $2.1 \times 1000 = 2100$.

Multiply to find the products.

1. $4.1 \times 100 =$
2. $3.7 \times 1000 =$
3. $7.4 \times 10 =$

4. $.4 \times 1000 =$
5. $6.555 \times 10 =$
6. $7.8 \times 100 =$

7. $2.5 \times 10 =$
8. $3.1 \times 100 =$
9. $64.72 \times 100 =$

10. $.563 \times 100 =$
11. $7.8 \times 1000 =$
12. $.74 \times 10 =$

13. $30.5 \times 10 =$
14. $42.123 \times 1000 =$
15. $.678 \times 100 =$

WORD PROBLEM

The key words *times* tells you to multiply to solve the following problem.

The distance from Laura's apartment to the office where she works is 1.25 miles. She walks that distance 10 **times** a week going to and from work. How far does she walk in a week between her home and her job?

Check your answers on page 232.

Lesson 27

Dividing Decimals and Mixed Decimals

Dividing decimals is about the same as dividing whole numbers. The only difference is that you have to place a decimal point correctly in the answer.

Dividing a Decimal or a Mixed Decimal by a Whole Number

Before you divide a decimal or a mixed decimal by a whole number, place a decimal point in the quotient directly above the decimal point in the dividend. Then divide as you would whole numbers. A zero is used to hold a decimal place in the quotient when you cannot divide.

Example 1: Divide .1236 by 4.

Step 1	Step 2	Step 3
$4\overline{)\,.1236}$	$4\overline{)\,.1236}$ (decimal point)	$4\overline{)\,.1236}$ with .0309 above

STEP 1: Set up the problem.

STEP 2: Place a decimal point in the quotient directly above the decimal point in .1236, the dividend.

STEP 3: Divide. Because you cannot divide 4 into the 1 in the tenths place of the dividend, write a 0 above the 1 to hold the place in the quotient. You also need to hold the thousandths place in the quotient with a 0.

EXERCISE 27a

Divide to find the quotients.

1. .8 ÷ 2 =
2. .6 ÷ 2 =
3. $.66 ÷ 6 =
4. .84 ÷ 4 =
5. $.93 ÷ 3 =
6. 3.9 ÷ 3 =
7. 8.4 ÷ 4 =
8. 7.28 ÷ 8 =
9. $4.86 ÷ 6 =
10. 2.48 ÷ 4 =
11. .028 ÷ 4 =
12. .012 ÷ 2 =
13. .024 ÷ 6 =
14. .063 ÷ 9 =
15. .021 ÷ 7 =
16. $3.66 ÷ 6 =
17. .684 ÷ 2 =
18. 1.12 ÷ 16 =
19. 59.86 ÷ 73 =
20. 127.3 ÷ 19 =

WORD PROBLEM

The key word *shared* tells you to divide to solve the following problem.

The cost of a taxi ride was **shared** by 2 people. The fare was $6.38. What did each person pay?

Check your answers on page 232.

Adding Zeros in the Dividend

Sometimes it is necessary to write one or more zeros to the right of a decimal or a mixed decimal in order to divide it. The zeros do not change the value of the decimal. They create places for the digits in the quotient.

Example 2: Divide .2 by 4.

Step 1	Step 2	Step 3
$4\overline{)\,.2}$	$4\overline{)\,.20}$	$4\overline{)\,.20}^{\,.05}$

STEP 1: Set up the problem and place a decimal point in the quotient directly above the decimal point in the dividend.

STEP 2: Because you cannot divide 4 into 2, write a zero to the right of the 2 to create a decimal place for the quotient.

STEP 3: Because you cannot divide 4 into 2, write a 0 above the 2. Divide 4 into 20 and write 5, the answer, above the 0 in the dividend.

EXERCISE 27b

Divide to find the quotients.

1. $.4 \div 5 =$ 2. $.3 \div 6 =$ 3. $.2 \div 5 =$ 4. $.3 \div 5 =$

5. $.1 \div 2 =$ 6. $.4 \div 8 =$ 7. $.6 \div 12 =$ 8. $.5 \div 10 =$

9. $.24 \div 48 =$ 10. $.36 \div 9 =$ 11. $.5 \div 10 =$ 12. $.6 \div 12 =$

13. $2.4 \div 60 =$ 14. $.236 \div 4 =$ 15. $.108 \div 12 =$ 16. $.075 \div 25 =$

WORD PROBLEM

The key words *equal parts* tell you to divide to solve the following problem.

A load of gravel that weighs .6 ton must be shipped in 8 **equal parts**. How much should be shipped in each part?

Check your answers on page 232.

As you know from Lesson 11, division problems do not always come out evenly. When you divide a decimal or a mixed decimal and the quotient does not come out evenly, write a zero to the right of the decimal in the dividend and divide again.

Example 3: Divide 3.9 by 4.

Step 1	Step 2	Step 3	Step 4

$$
\begin{array}{r} . \\ 4\overline{)3.9} \end{array}
\qquad
\begin{array}{r} .9 \\ 4\overline{)3.9} \\ \underline{3\,6} \\ 3 \end{array}
\qquad
\begin{array}{r} .97 \\ 4\overline{)3.90} \\ \underline{3\,6} \\ 30 \\ \underline{28} \\ 2 \end{array}
\qquad
\begin{array}{r} .975 \\ 4\overline{)3.900} \\ \underline{3\,6} \\ 30 \\ \underline{28} \\ 20 \\ \underline{20} \\ 0 \end{array}
$$

STEP 1: Set up the problem and place a decimal point in the quotient directly above the decimal point in the dividend.

STEP 2: Because you cannot divide 4 into 3, divide it into 39. Write the 9 in the quotient. Multiply: $9 \times 4 = 36$. Write the 36 under the 39. Subtract: $39 - 36 = 3$. The quotient does not come out even.

STEP 3: Write a zero to the right of the mixed decimal in the dividend: 3.9**0**. Bring the zero down and divide: $30 \div 4 = 7$. Write the 7 in the quotient. Multiply: $7 \times 4 = 28$. Write the 28 under the 30. Subtract: $30 - 28 = 2$. The quotient still does not come out even.

STEP 4: Write another zero to the right of the mixed decimal in the dividend: 3.9**00**. Bring down the zero and divide: $20 \div 4 = 5$. Write the 5 in the quotient. Multiply: $5 \times 4 = 20$. Write the 20 under the 20 and subtract: $20 - 20 = 0$. The quotient comes out even. It is .975.

EXERCISE 27c

Divide to find the quotients.

1. $2.8 \div 8 =$ 2. $.38 \div 5 =$ 3. $2.6 \div 4 =$ 4. $.51 \div 6 =$

5. $4.1 \div 2 =$ 6. $3.1 \div 5 =$ 7. $.3 \div 24 =$ 8. $.3 \div 8 =$

9. $1.5 \div 6 =$ 10. $.6 \div 5 =$ 11. $1.407 \div 6 =$ 12. $.1 \div 16 =$

13. $7.8 \div 8 =$ 14. $.8 \div 5 =$ 15. $.7 \div 4 =$ 16. $.16 \div 5 =$

The key word *evenly* tells you that you need to divide to solve the following problem.

Oscar wants to divide 4.5 pounds of ground beef into 4 portions. If he divides the ground beef **evenly**, how much will each portion weigh?

Check your answers on page 232.

Rounding Quotients

Some division problems never come out evenly. Some don't come out evenly until several zeros are added to the right of the decimal in the dividend. In most cases like this, unless you are instructed otherwise, you should divide to the ten-thousandths place (the fourth decimal place) and then round the quotient to the thousandths place (the third decimal place). That will give you an answer with three decimal places.

Example 4: Divide .6 by 7.

Step 1	Step 2	Step 3	Step 4
.08	.085	.0857	
7)‾.60	7)‾.600	7)‾.6000	.086
56	56	56	
4	40	40	
	35	35	
	5	50	
		49	
		1	

STEP 1: Set up the problem and place a decimal point in the quotient directly above the decimal point in the dividend. Because you cannot divide 7 into 6, write a zero in the quotient. Write a zero to the right of the dividend: .60. Divide: $60 \div 7 = 8$. Multiply: $8 \times 7 = 56$. Write the 56 under the 60. Subtract: $60 - 56 = 4$. The quotient does not come out even.

STEP 2: Write another zero to the right of the decimal in the dividend: 6.00. Bring the zero down and divide: $40 \div 7 = 5$. Write the 5 in the quotient. Multiply: $5 \times 7 = 35$. Write the 35 under the 40. Subtract: $40 - 35 = 5$. The quotient still does not come out even.

STEP 3: Write a third zero to the right of the mixed decimal in the dividend: .6000. Bring down the zero and divide: $50 \div 7 = 7$. Write the 7 in the quotient. Multiply: $7 \times 7 = 49$. Write the 49 under the 50. The quotient still has not come out even, but you have divided to the ten-thousandths place.

STEP 4: Round the quotient to the nearest thousandth. Because the 7 in the ten-thousandths place is larger than 5, change the 5 to a 6 and drop the 7. The answer is .086.

Part A. Divide. Round each quotient to the nearest tenth.

1. $2.8 \div 6 =$ 2. $1.38 \div 5 =$ 3. $.28 \div 6 =$

4. $4.3 \div 7 =$ 5. $.095 \div 8 =$

Part B. Divide. Round each quotient to the nearest hundredth.

1. $3.7 \div 20 =$ 2. $.137 \div 2 =$ 3. $2.27 \div 7 =$

4. $4.23 \div 35 =$ 5. $1.208 \div 4 =$

Part C. Divide. Round each quotient to the nearest thousandth.

1. $8.1 \div 7 =$ 2. $49.3 \div 6 =$ 3. $2.5 \div 6 =$

4. $11.5 \div 3 =$ 5. $2.51 \div 8 =$

WORD PROBLEM

The key words *divided* and *evenly* tell you to divide to solve the following problem.

The 7 employees in the shipping department **divided** the cost of a $60.95 coffeemaker **evenly** among themselves. Rounded to the nearest cent, how much did each employee contribute to the cost of the coffeemaker?

Check your answers on page 232.

Dividing by a Decimal or a Mixed Decimal

To divide by a decimal or a mixed decimal, first move the decimal point in the divisor all the way to the right. Then move the decimal point in the dividend the same number of places to the right. After that, divide as usual. You will be dividing by a whole number.

Before you work a problem, estimating the answer can help you be sure that the answer you get when you work the problem makes sense. Your estimate and the answer should be close to each other in value.

Example 5: Divide 4.5 by 2.5.

Step 1	Step 2
$2.5.\overline{)4.5.}$	$\begin{array}{r} 1.8 \\ 2.5.\overline{)4.5.0} \\ \underline{2\ 5} \\ 2\ 0\ 0 \\ \underline{2\ 0\ 0} \\ 0 \end{array}$
Estimate:	
$40 \div 20 = 2$	

STEP 1: After you set up the problem, move the decimal point in 2.5 all the way to the right: 2.5 becomes 25. Because you moved the decimal point in the divisor one place, move the decimal point in the dividend one place: 4.5 becomes 45. You can estimate the answer by dividing: $40 \div 20 = 2$.

STEP 2: Place a decimal point in the quotient above the decimal point in 45. Divide as usual: $45 \div 25 = 1.8$. Compare the answer with your estimate in Step 1: they are close in value.

EXERCISE 27e

Divide to find the quotients.

1. $.8 \div .1 =$

2. $4.2 \div .6 =$

3. $.81 \div .9 =$

4. $5.6 \div .7 =$

5. $.48 \div .12 =$

6. $.406 \div .7 =$

7. $.424 \div .8 =$

8. $7.25 \div .05 =$

9. $40.2 \div 6.7 =$

10. $138.84 \div 3.9 =$

11. $38.4 \div 1.28 =$

12. $6.7 \div 1.34 =$

13. $91.5 \div .125 =$

14. $457.8 \div 32.7 =$

15. $3.9 \div .013 =$

16. $3.9 \div 1.3 =$

17. $91.3715 \div 4.07 =$

18. $77.5 \div .31 =$

19. $84.6 \div 5.64 =$

20. $25.27 \div .722 =$

WORD PROBLEM

The key word *each* tells you to divide to solve the following problem.

Luz packed 57.5 pounds of apples into boxes. She packed 2.5 pounds into **each** box. How many boxes did Luz pack?

Check your answers on page 232.

Dividing a Whole Number by a Decimal or a Mixed Decimal

When you divide a whole number by a decimal or a mixed decimal, first write a decimal at the end of the whole number. Then move the decimal points in both the divisor and the dividend. After than, divide as usual. You will be dividing by a whole number.

Before you work a problem, estimate. When you have worked the problem, check your answer against your estimate. They should be close to each other in value.

Example 6: Divide 576 by 1.75.

Step 1	Step 2	Step 3	Step 4

$$1.75\overline{)567.}$$ $$1.75.\overline{)567.00.}$$ $$1.75.\overline{)567.00.}$$

$$\begin{array}{r} 324. \\ 175\overline{)56700.} \\ 525 \\ \hline 420 \\ 350 \\ \hline 700 \\ 700 \\ \hline 0 \end{array}$$

Estimate:
$560 \div 2 = 280$

STEP 1: Set up the problem. Write a decimal point at the end of the whole number.

STEP 2: Move the decimal point in 1.75 all the way to the right: 1.75 becomes 175. Because you moved the decimal point in the divisor two places, move the decimal point in the dividend two places: 567 becomes 56700.

STEP 3: Place a decimal point in the quotient above the decimal point in 56700.

STEP 4: Divide as usual: $56700 \div 175 = 324$.

EXERCISE 27f

Divide to find the quotients.

1. $56 \div .7 =$

2. $30 \div .6 =$

3. $81 \div .9 =$

4. $40 \div .8 =$

5. $49 \div .7 =$

6. $12 \div .4 =$

7. $1 \div .5 =$

8. $12 \div .04 =$

9. $15 \div .003 =$

10. $35 \div .007 =$

11. $96 \div 1.2 =$

12. $69 \div .3 =$

13. $32 \div .04 =$

14. $64 \div .008 =$

15. $48 \div .006 =$

16. $780 \div 1.3 =$

17. $2368 \div .32 =$

18. $560 \div .8 =$

19. $34,848 \div .0066 =$

20. $4578 \div 3.27 =$

WORD PROBLEM

The key word *each* tells you to divide to solve the following problem.

Harry packed 480 pounds of shrimp into packages that weighed .8 pound **each**. How many packages did Harry pack?

Check your answers on page 233.

Dividing Decimals and Mixed Decimals by 10, 100, and 1000

To divide a decimal or mixed decimal by 10, 100, or 1000, just move the decimal point one place to the left for every zero in the number you are dividing by.

Example 7: Divide 21.234 by 10, by 100, and by 1000.

$$21.234 \div 10 = 2.1234$$
$$21.234 \div 100 = .21234$$
$$21.234 \div 1000 = .021234$$

As the third example shows, it is sometimes necessary to write a zero at the left end of a decimal in order to move the decimal point enough places.

EXERCISE 27g

Divide to find the quotients.

1. $3.2 \div 100 =$
2. $4.5 \div 1000 =$
3. $.6 \div 10 =$

4. $.9 \div 100 =$
5. $.04 \div 1000 =$
6. $.89 \div 10 =$

7. $3.6 \div 1000 =$
8. $13.20 \div 100 =$
9. $.9 \div 1000 =$

10. $1.2 \div 100 =$
11. $26.4 \div 10 =$
12. $11.8 \div 1000 =$

13. $6.816 \div 10 =$
14. $.67 \div 100 =$
15. $.834 \div 1000 =$

16. $.23 \div 10 =$

WORD PROBLEM

The key words *equal* and *each* tell you to divide to solve the following problem.

Angelica sawed a board 16.4 feet long into 10 **equal** pieces. How many feet long was **each** piece?

Check your answers on page 233.

MIXED PRACTICE 4
MULTIPLICATION AND DIVISION OF DECIMALS

These problems let you practice the decimal multiplication and division skills you have learned. Read each problem carefully and solve it.

1. $.30 × 6 =

2. .87 × .26 =

3. .07 × 4 =

4. .048 × 5 =

5. 6.75 × 2.4 =

6. .12 ÷ 4 =

7. 135 ÷ .45 =

8. .78 ÷ .6 =

9. 8.47 ÷ 1000 =

10. 13.5 ÷ 5 =

11. 1.77 × 6 =

12. 4.27 × 5 =

13. 9.836 × .02 =

14. 32.59 × 4 =

15. 56.06 × 7 =

16. 2 ÷ .25 =

17. 1.38 ÷ .06 =

18. 95.4 ÷ 6.36 =

19. 444 ÷ 3.7 =

20. 27.6 ÷ .4 =

21. 37.1 × .25 =

22. 41.8 × .75 =

23. 84.6 × 12 =

24. $35.65 × 2 =

25. $1.39 × 74 =

26. 14.4 ÷ 2.4 =

27. 1.35 ÷ .54 =

28. .006 ÷ .4 =

29. 5.6 ÷ 1.12 =

30. 7.2 ÷ .9 =

31. .316 × .93 =

32. 41.8 × 7.5 =

33. $.59 × 25 =

34. .6001 × 206 =

35. 1.5 × .13 =

36. 7.24 ÷ 100 =

37. 95.4 ÷ .06 =

38. 12.92 ÷ .323 =

39. 130.66 ÷ 4.7 =

40. 83.6 ÷ 10 =

41. $6.02 × 82 =

42. .36 × .22 =

43. $64.98 × 41 =

44. .88 × 306 =

45. .18 × .21 =

46. $1.28 ÷ 32 =

47. $17.50 ÷ 14 =

48. 200.50 ÷ 25 =

49. 246.42 ÷ 6 =

50. 1.2 ÷ .08 =

Check your answers on page 233.

DECIMAL SKILLS REVIEW

Part A. Add, subtract, multiply, or divide.

1. 3.53 + .457 =
2. $101.99 − $1.99 =
3. .341 × 6 =
4. 18.99 ÷ 9 =
5. 1.3 + .04 + .637 =
6. 16.237 − 8.126 =
7. $7.25 × 6 =
8. .1505 ÷ 5 =
9. 4.39 + 2.46 =
10. 10.21 − 8.1 =
11. .638 × .45 =
12. $78.06 + $16.79 =
13. 33.176 − 32.15 =
14. $.45 × .5 =
15. 11.75 ÷ 5 =
16. 3.224 ÷ 8 =
17. 325.11 + 49.79 =
18. 645.8 − 51.672 =
19. 3.712 × .04 =
20. .6 ÷ 12 =
21. 6.21 + .99 =
22. 37.832 − 6.843 =
23. 81.45 × 8.7 =
24. .02 ÷ 4 =
25. 18 + .64 + 1.21 =
26. $46 − $.57 =
27. 38.4 ÷ 1000 =
28. 16 − .001 =
29. 236.4 × .09 =
30. 1.407 ÷ 6 =
31. 11.5 ÷ 3 =
32. .09 × 1.06 =
33. 45.71 ÷ .7 =
34. 9.145 ÷ 2.95 =
35. 2.4 × 10 =
36. 39.6 ÷ .036 =
37. 36 ÷ .06 =
38. .16 × 1000 =
39. 2300 ÷ 1.15 =
40. 1 + .1 + 1.1 + .01 + 10 =

Part B. Solve each problem.

1. Alex filled his gas tank at the service station. It held ten and twelve hundredths gallons. The gas pump showed the amount in a mixed decimal. What did the pump show?

2. At the deli, Mallory bought .75 pound of sliced ham, 1.2 pounds of smoked fish, and .8 pound of potato salad. Which of his three purchases weighed the most? Which weighed the least?

3. Mariama weighed 5 items on a precision scale. The items weighed as follows: (a) .501 lb, (b) .155 lb, (c) .51 lb, (d) 1.01 lb, and (e) .015 lb. List the 5 items in order of size from smallest to largest.

4. George wanted to report each of these measurements rounded to the nearest tenth: (a) 7.129 lb, (b) 25.654 km, (c) 18.009 g, and (d) .08 cm. What should George have reported for each measurement?

5. On Monday it rained 1.09 inches; on Tuesday, .88 inch; on Wednesday, 1.13 inches; and on Thursday, .2 inch. What was the total rainfall for the 4 days?

6. A certain part in one computer weighs .43 g. In another computer, a similar part weighs .396 g. How much more does the heavier part weigh than the lighter one?

7. Rounded to the nearest cent, what is the total cost of .8 oz of a spice that costs $.07 per ounce?

8. Altogether, 9 seeds weigh .171 oz. How much does each seed weigh?

Check your answers on page 233.

Chapter

4 USING DECIMALS

In the seven lessons in this chapter, you will solve word problems and problems that involve metric measurements. You will find averages; calculate the perimeter, circumference, and area of figures; use the cost formula and find unit costs; and learn to solve multistep problems.

Lesson 28

Solving Decimal Word Problems

The steps for solving decimal word problems are the same as those for solving whole-number word problems in Lesson 18:

- Understand the question.
- Find the facts you need to solve the problem.
- Decide which math operation to use.
- Estimate the solution.
- Solve the problem and check your answer.

Before you begin the following exercise, you may wish to review Lesson 18.

EXERCISE 28

Read each problem carefully, making sure you understand the question. Find the facts you need to solve the problem and decide which operation to use. Estimate the solution, and then solve the problem. Check your answer.

1. Rita earns $5.35 per hour at a part-time job. She worked 18 hours this week. How much did she earn in all?

2. In 1941 Ted Williams had a batting average of .406. The next highest batting average was Pete Reiser's .343. How many points higher was Ted William's batting average?

3. Marge has an adjustable wrench. She opens it .6 inch to tighten a nut. It's too small, so she opens it .2 inch more. How far has she opened the wrench altogether?

4. A lot of chocolate that weighs 360 pounds is packed in 1.25-pound boxes. How many boxes of chocolate are packed?

5. Leo jogs 3.25 miles 7 times a week. How many miles does he jog in one week?

6. Hector drove his stock car around a track in 45.04 seconds. Luis drove his car around the same track in 43.76 seconds. How many seconds less did Luis take to drive around the track?

7. Alicia wrote the following checks: $300 for rent, $45.76 for electricity, and $167.89 for her car payment. What was the total amount of the checks Alicia wrote?

8. Carlos makes the 35.6-mile trip to and from his office 10 times each week. What is the total distance Carlos drives between home and work each week?

9. A stack of plywood sheets is 60 inches high. Each sheet is 1.5 inches thick. How many sheets are in the stack?

10. At her job, Pauline had to find the total weight of 1076 steel bolts. She found that one bolt weighed .75 ounce. How many ounces did all the bolts weigh?

Check your answers on page 233.

Lesson 29

Metric Measurements

In Lesson 13 you learned about the English and the metric systems of measurement. You changed large units of measure into smaller units of measure, and vice versa.

The metric system is based on the decimal system. Using decimal skills makes changing between large and small units of measure in the metric system easier.

Metric Relationships

The following chart shows how small and large units of measure are related to each other in the metric system.

The top row of the chart shows the six prefixes that are used to describe measurements: kilo, hecto, deka, deci, centi, and milli. The second row of the chart shows how the values associated with each prefix are related to the basic unit of measure. The last three rows of the chart give the names of the various units used to measure length, liquid, and weight.

Prefix	kilo	hecto	deka		deci	centi	milli
Value	Basic Unit × 1000	Basic Unit × 100	Basic Unit × 10	**Basic Unit**	Basic Unit × 0.1	Basic Unit × 0.01	Basic Unit × 0.001
Length	kilometer (km)	hectometer (hm)	dekameter (dam)	meter (m)	decimeter (dm)	centimeter (cm)	millimeter (mm)
Liquid	kiloliter (kL)	hectoliter (hL)	dekaliter (daL)	liter (L)	deciliter (dL)	centiliter (cL)	milliliter (mL)
Weight	kilogram (kg)	hectogram (hg)	dekagram (dag)	gram (g)	decigram (dg)	centigram (cg)	milligram (mg)

By reading the chart, you can see how one unit of measure relates in size to another.

Example 1: Compared with a meter, how long is a kilometer?

STEP 1: Find *meter* on the chart. It is the basic unit of measure of length.

STEP 2: Find *kilometer* on the chart. It is in the column under the prefix *kilo*. As the chart shows, a measurement that begins with the prefix *kilo* equals the basic unit multiplied by 1000. Therefore, a kilometer equals 1 meter × 1000, or 1000 meters.

EXERCISE 29a

Answer each of the following questions.

1. Compared with a liter, how large is a milliliter?
2. Compared with a gram, how large is a hectogram?
3. Compared with a meter, how long is a centimeter?
4. Compared with a liter, how large is a deciliter?
5. Compared with a gram, how large is a dekagram?

Check your answers on page 234.

Changing from One Unit of Measure to Another

In the metric system, to change from one unit of measure to another, all you need to do is move the decimal point to the right or to the left.

Example 2: Change 22.5 dekameters to meters.

dekameters		meters
22.5	=	225.

On the chart, a meter is one place to the right of a dekameter. Therefore, to change dekameters to meters, just move the decimal point one place to the right. 22.5 dekameters equals 225 meters.

EXERCISE 29b

Change each unit of measure to the one shown.

1. 2.5 kg = _____ g

2. 1.2 daL = _____ L

3. 2 hm = _____ m

4. 200 dm = _____ m

5. 40 mm = _____ cm

6. .2 cm = _____ m

7. 4.5 L = _____ mL

8. 5.03 km = _____ mm

9. 960 g = _____ kg

10. 344.8 L = _____ kL

WORD PROBLEM

A baker has 2.5 kilograms of raisins. How many grams of raisins is that?

Check your answers on page 234.

Check your answers on page 234.

Lesson 30

Finding Averages

In Lesson 17 you found averages of sets of whole numbers. Sometimes you need to find the average of a set of decimals and/or mixed decimals. The method for doing that is the same as the method you used with whole numbers: to find an average, divide the sum of a set of numbers by the number of numbers in the set.

Example: Find the average of this set of numbers: 8.5, .98, 84, 7.5, and 1.93.

Step 1	Step 2

Step 1:

```
    8.5
    .98
   84.
    7.5
 + 1.93
 ─────────
 102.91
```

Step 2:

```
        20.582
    5)102.910
      10
      ──
      02 9
       2 5
       ───
         41
         40
         ──
         10
         10
         ──
          0
```

STEP 1: Add to find the sum of the five numbers in the set.

STEP 2: Divide the sum by 5, the number of numbers in the set. The average of this set of numbers is 20.582.

EXERCISE 30

Find the average of each set of numbers. Round money answers to the nearest cent. As it is necessary, round other answers to the nearest ten thousandth.

1. 1.2, 1.5, 2.5, 8

2. $2.31, $3.21, $1.32

3. .123, 5.8, .72, 96, 6.5, 54, .45

4. 4.1, .42, .043, .044, .145, 4.06

5. 700.65, 900.85, 930

6. .48, .52, .48, .39, .57, .5

7. 2.54, 3.8, 9.4, 4.8, 7.815

8. $.22, $.30, $.18, $.21, $.34, $3.09

9. 2.4 m, 1.2 m, 2.28 m, 1.44 m, 8.4 m, 17.7 m

10. 2.3 km, 5.2 km, 1.8 km, 5.5 km, 2.4 km, 10 km, .4 km, .4 km, 6.5 km, 6 km

WORD PROBLEM

Andy had a fever for 5 days. He took his temperature each morning. These were the readings: 99.2, 100.6, 101.1, 99.8, 99.1. What was the average of the readings, rounded to the nearest tenth?

Check your answers on page 234.

Finding Perimeter and Area
with Metric Measurements

As you read in Lesson 15, triangles, squares, and rectangles are closed figures with straight sides. The perimeter is the distance around such a straight-sided figure. The area is the amount of surface area on a flat figure.

A figure's measurements can be given with the metric system. To find its perimeter or area, you use the same methods you practiced in Lesson 15. The only difference is that you may need to add, multiply, or divide decimals, since metric measurements are often expressed as mixed decimals.

Perimeter

To find the perimeter of a figure, add the lengths of all its sides. Remember: To find the perimeter of a square, it is faster to multiply, since all the sides are the same length. The same is true for triangles with sides of equal length.

Example 1: Find the perimeter of each of the following figures.

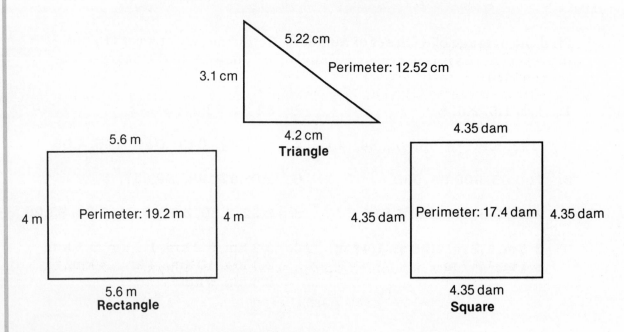

Triangle: To find the perimeter, add:

$$3.1 \text{ cm} + 4.2 \text{ cm} + 5.22 \text{ cm} = 12.52 \text{ cm}$$

Rectangle: To find the perimeter, add:

$$4 \text{ m} + 5.6 \text{ m} + 4 \text{ m} + 5.6 \text{ m} = 19.2 \text{ m}$$

Square: To find the perimeter, multiply:

$$4.35 \text{ dam} \times 4 = 17.4 \text{ dam}$$

Find the perimeter of each of these figures.

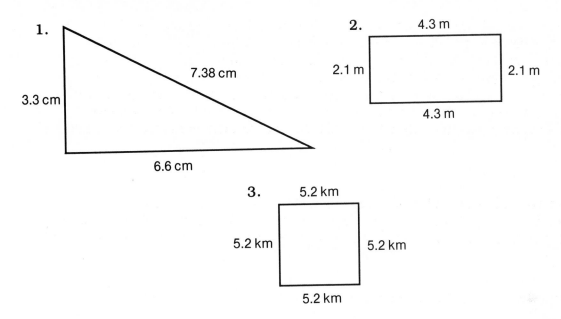

1.

7.38 cm

3.3 cm

6.6 cm

2.   4.3 m

2.1 m   2.1 m

4.3 m

3.   5.2 km

5.2 km   5.2 km

5.2 km

WORD PROBLEM

A rectangular garden measures 3.5 meters by 4 meters. How many meters of fencing are needed to enclose the garden?

Check your answers on page 234.

Area

To find the area of a four-sided figure, multiply its length by its width:

$$\text{Area} = \text{length} \times \text{width}.$$

To find the area of a triangle, multiply its base by its height and divide by 2:

$$\text{Area} = \frac{\text{base} \times \text{height}}{2}$$

Remember that area is expressed in square units of measure.

When you multiply metric measurements, there may be several decimal places in the answer. Unless you are told otherwise, round the answer to the nearest thousandth.

Example 2: Find the area of each of the following figures.

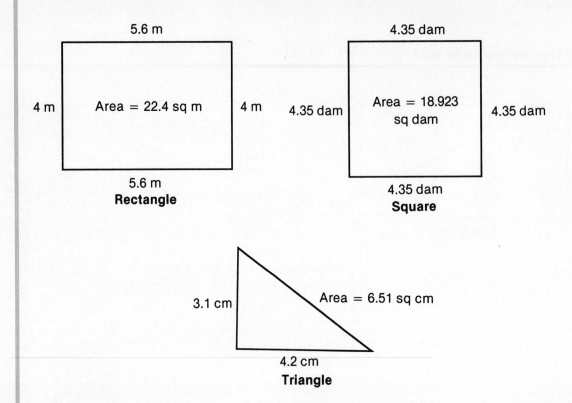

Rectangle: To find the area, use this formula:

$$\text{Area} = \text{length} \times \text{width}$$
$$\text{Area} = 5.6 \text{ m} \times 4 \text{ m}$$
$$\text{Area} = 22.4 \text{ sq m}$$

Square: To find the area, use this formula:

$$\text{Area} = \text{length} \times \text{width}$$
$$\text{Area} = 4.35 \text{ dam} \times 4.35 \text{ dam}$$
$$\text{Area} = 18.923 \text{ sq dam}$$

Triangle: To find the area, use this formula:

$$\text{Area} = \frac{\text{base} \times \text{height}}{2}$$
$$\text{Area} = \frac{4.2 \text{ cm} \times 3.1 \text{ cm}}{2}$$
$$\text{Area} = 6.51 \text{ sq cm}$$

Use the appropriate formula to find the area of each of the following figures.

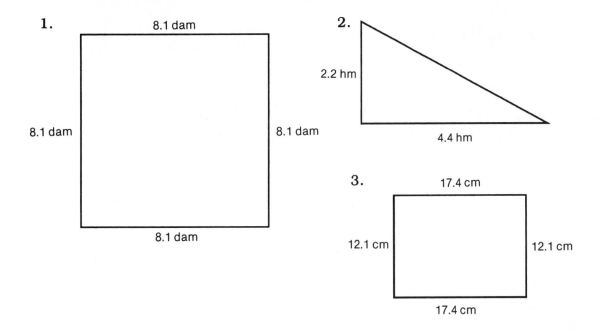

1.
8.1 dam
8.1 dam
8.1 dam
8.1 dam

2.
2.2 hm
4.4 hm

3.
17.4 cm
12.1 cm
12.1 cm
17.4 cm

WORD PROBLEM

A rectangular kitchen measures 3.5 meters by 4 meters. How many square meters of linoleum are needed to cover the floor?

Check your answers on page 234.

For more practice in finding the perimeter and area using metric measurements, do the following exercise.

EXERCISE 31c

Find both the perimeter and the area of each of the following figures.

1.
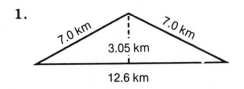
7.0 km   7.0 km
3.05 km
12.6 km

2.
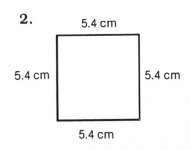
5.4 cm
5.4 cm
5.4 cm
5.4 cm

3.

4.6 mm   4.6 mm
3.99 mm
4.6 mm

4.

4.2 dm
4.2 dm   4.2 dm
4.2 dm

5.

8 km
14.3 km   14.3 km
8 km

6.

6.5 hm
11.7 hm   11.7 hm
6.5 hm

7.

4.4 hm   4.4 hm
3.8 hm
4.4 hm

8.

12.6 dam
12.6 dam   12.6 dam
12.6 dam

9.

4.4 m
2 m   2 m
4.4 m

Check your answers on page 234.

Finding the Circumference and Area of Circles

A circle is a curved, flat figure with every point on it an equal distance from
its center. To find the circumference or the area of a circle, you must know
the relationships among its parts.

The Parts of a Circle

The **circumference** of a circle is the distance around the circle. A line
that runs from one side of a circle, through the center, to the opposite side
is the **diameter**. The **radius** of a circle is a line that runs from the center to
the edge.

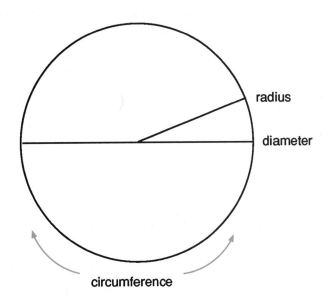

The diameter of a circle is two times as long as its radius. Therefore,
the radius of a circle is half as long as its diameter.

Example 1: Find the missing measurement for each of the following circles.

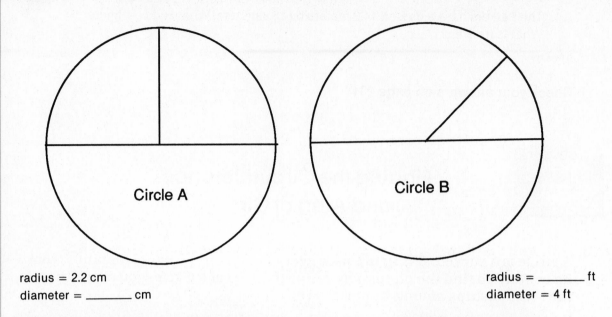

radius = 2.2 cm
diameter = _____ cm

radius = _____ ft
diameter = 4 ft

Circle A: The radius is 2.2 centimeters. Therefore, the diameter, which is two times as long as the radius, is 4.4 centimeters.

Circle B: The diameter is 4 feet. Therefore, the radius, which is half as long as the diameter, is 2 feet.

EXERCISE 32a

Find the missing measurement for each of the circles described here.

1. radius   = 15 ft

   diameter = _____

2. diameter = 1.25 cm

   radius   = _____

3. radius   = 15.75 cm

   diameter = _____

4. radius   = 200.3 dm

   diameter = _____

5. diameter = 13.6 cm

   radius   = _____

6. radius   = 49 in.

   diameter = _____

7. diameter = 7.5 km

   radius   = _____

8. diameter = 66 yd

   radius   = _____

9. diameter = 19.4 m

   radius   = _____

10. radius   = 23.5 mi

    diameter = _____

Check your answers on page 235.

Finding the Circumference of a Circle

The circumference of a circle is about 3.14 times as long as the circle's diameter. The following diagram shows that relationship.

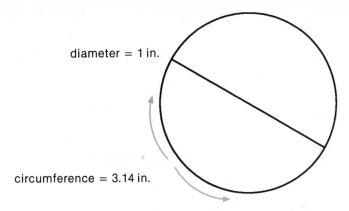

The relationship between the diameter and the circumference of a circle can be expressed by this formula:

Circumference = 3.14 × diameter

Using this formula, you can find the circumference of any circle.

Example 2: Find the circumference of a circle with a diameter of 4 inches.

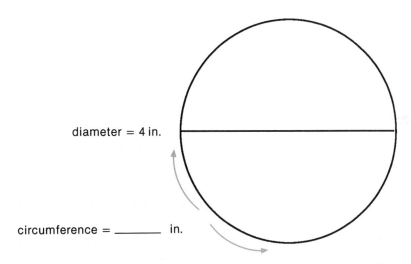

Step 1	**Step 2**
Circumference = 3.14 × diameter	Circumference = 3.14 × diameter
	Circumference = 3.14 × 4 inches
	Circumference = 12.56 inches

STEP 1: Write the formula for the circumference of a circle.

STEP 2: Rewrite the formula, substituting *4 inches* for the word *diameter*. Then, multiply to find the circumference, which is 12.56 inches.

Find the circumference of each of the following circles. Round your answers to the nearest thousandth.

1. diameter = 6 in.

2. diameter = 10 ft

3. diameter = 1.6 m

4. diameter = .8 cm

5. diameter = 325 ft

6. diameter = 40 yd

7. diameter = 64 mi

8. diameter = .18 m

9. diameter = 1.1 mm

10. diameter = 10.5 km

Check your answers on page 235.

Sometimes when you want to find the circumference of a circle, you do not know its diameter, but you know its radius. Before you can find the circumference, you must find the diameter.

Example 3: Find the circumference of a circle with a radius of 4 meters.

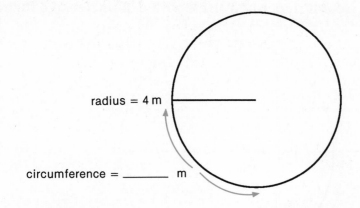

radius = 4 m

circumference = _____ m

Step 1	**Step 2**
The radius = 4 meters; therefore the diameter = 8 meters.	Circumference = 3.14 × diameter Circumference = 3.14 × 8 meters Circumference = 25.12 meters

STEP 1: To find the diameter, double the radius: 4 meters × 2 = 8 meters.

STEP 2: Write the formula for the circumference of a circle. Rewrite the formula, substituting *8 meters* for the word *diameter*. Then, multiply to find the circumference, which is 25.23 meters.

Find the circumference of each of the following circles. Round your answers to the nearest thousandth.

1. radius = .6 m
2. radius = 10 ft
3. radius = 10.5 m
4. radius = .8 cm
5. radius = 3.21 dam
6. radius = 40 yd
7. radius = 64 mi
8. radius = .18 m
9. radius = 1.1 mm
10. radius = 1.5 km

Check your answers on page 235.

Finding the Area of a Circle

There is also a formula for finding the area of a circle:

$$\text{Area} = 3.14 \times \text{radius} \times \text{radius}$$

By using this formula, you can find the area of any circle. Remember that area is expressed in square units of measure.

Example 4: Find the area of a circle with a radius of 6 yards.

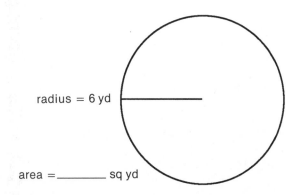

radius = 6 yd

area = _____ sq yd

Step 1

Area = 3.14 × radius × radius

Step 2

Area = 3.14 × radius × radius
Area = 3.14 × 6 yards × 6 yards
Area = 113.04 square yards

STEP 1: Write the formula for the area of a circle.

STEP 2: Rewrite the formula, substituting *6 yards* for the word *radius.* Then, multiply to find the area, which is 113.04 square yards.

Find the area of each of the following circles. Round your answers to the nearest thousandth.

1. radius = 14 in.

2. radius = 1.5 cm

3. radius = 2.3 mi

4. radius = 1 m

5. radius = 44 yd

6. radius = .85 km

7. radius = 13 in.

8. radius = .05 m

9. radius = 1.1 mm

10. radius = 10.5 km

Check your answers on page 236.

Sometimes when you want to find the area of a circle, you do not know its radius, but you know its diameter. Before you can find the area, you must find the radius.

Example 5: Find the area of a circle with a diameter of 4 meters.

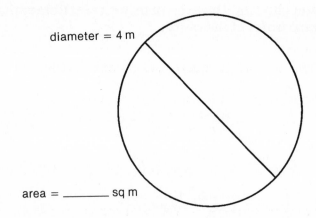

diameter = 4 m

area = _____ sq m

Step 1

The diameter = 4 meters; therefore the radius = 2 meters.

Step 2

Area = 3.14 × radius × radius
Area = 3.14 × 2 meters × 2 meters
Area = 12.56 square meters

STEP 1: To find the radius, divide the diameter by 2:
4 meters ÷ 2 = 2 meters.

STEP 2: Write the formula for the area of a circle. Rewrite the formula substituting *2 meters* for the word *radius*. Then, multiply to find the area, which is 12.56 meters.

Find the area of each of the following circles. Round your answers to the nearest thousandth.

1. diameter = 6 in.

2. diameter = 10 ft

3. diameter = 1.6 m

4. diameter = .8 cm

5. diameter = 324 ft

6. diameter = 40 yd

7. diameter = 64 mi

8. diameter = .18 m

9. diameter = 1.12 mm

10. diameter = 10.5 km

WORD PROBLEM

To lay out a circular garden, Carolyn tied a rope 3 yards long to a stake, stretched it taut, and drew a circle, as the following diagram shows.

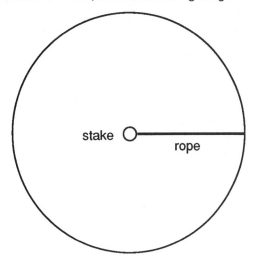

What were the radius, the diameter, the circumference, and the area of Carolyn's garden?

Check your answers on page 236.

Using the Cost Formula

You probably make frequent use of decimals when you are shopping. To find the total cost for more than one of an item, you multiply. To find the unit cost of an item, you divide.

Finding the Total Cost

Suppose you are buying cherries. You know the price of 1 pound, but you want to find the total cost for 3 pounds. To find the total cost, you need to multiply. The method for finding the total cost can be expressed in a formula:

$$\text{Total cost} = \text{number of units} \times \text{rate per unit}$$

The cost formula can be simplified:

$$\text{Cost} = \text{number} \times \text{rate}$$

Example 1:   Find the cost of 3 pounds of cherries at $1.29 per pound.

Step 1	**Step 2**	**Step 3**
Cost = number × rate	Cost = number × rate	Cost = number × rate
	Cost = 3 lb × $1.29	Cost = 3 lb × $1.29
		Cost = $3.87

STEP 1: Write the formula for finding the total cost.

STEP 2: Rewrite the formula, substituting *3 pounds* and *$1.29* for the words *number* and *rate.*

STEP 3: Multiply 3 pounds by $1.29. Rewrite the formula to state the total cost, $3.87.

Whenever you use the cost formula, it is a good idea to estimate your answer before you work the problem. That way you will know if your answer is reasonable. For instance, with the example above, you could have estimated this way: 3 × $1.30 = $3.90. With that estimate, you know that the answer, $3.87, is probably correct. If your answer had been $38.70, you would have known that it was probably wrong because it is too much greater than the estimate.

EXERCISE 33a

Use the cost formula to find the total cost of each of the following purchases. Round your answers to the nearest cent.

1. 5.67 pounds of chicken at $1.39 per pound.

2. 1.3 pounds of cheddar cheese at $2.49 per pound.

3. 52 feet of wood planking at $.39 per running foot.

4. 4.75 pounds of beef at $3.49 per pound.

5. 250 bricks at $.43 each.

6. 7.5 yards of canvas at $1.30 per yard.

7. 6 pairs of socks at $3.25 a pair.

8. 20 bus tokens at $1.15 each.

9. 1.4 pounds of walnuts at $2.58 per pound.

10. 4 sweaters at $34.75 each.

Check your answers on page 236.

Finding the Unit Cost

In a grocery store, a 24-ounce can of something may cost $1.75. To find out how much each unit, or ounce, of the product costs, you can use the unit-cost formula. The method for finding unit cost can be expressed in a formula:

$$\text{Unit cost} = \text{price} \div \text{number of units}$$

The unit-cost formula can be simplified:

$$\text{Unit cost} = \text{price} \div \text{units}$$

Example 2: Find the unit cost of a soup whose price is $.72 for 6 ounces.

Step 1	**Step 2**	**Step 3**
Unit cost = price ÷ units	Unit cost = price ÷ units	Unit cost = price ÷ units
	Unit cost = $.72 ÷ 6 oz	Unit cost = $.72 ÷ 6 oz
		Unit cost = $.12 per oz

STEP 1: Write the formula for finding the unit cost.

STEP 2: Rewrite the formula, substituting *$.72* and *6 ounces* for the words *price* and *units*.

STEP 3: Divide $.72 by 6 ounces. Rewrite the formula to state the unit cost, $.12 per ounce.

As with the cost formula, it is a good idea to estimate your answer before you work a unit-cost problem. With the example above, you could have estimated this way: $.70 ÷ 7 = $.10. With that estimate, you know that the answer, $.12, is probably correct. If your answer had been $1.20, you would have known that it was probably wrong because it is too much greater than the estimate.

EXERCISE 33b

Use the formula to find the unit cost of each of the following products. Round your answers to the nearest cent.

1. The cost per pound for bread that sells at $.99 for 1.2 pounds.

2. The cost per pound for cheese that sells at $1.79 for .75 pound.

3. The cost per ounce for tomato sauce that sells at $.81 for 18 ounces.

4. The cost per ounce for tuna fish that sells at $1.29 for 6 ounces.

5. The cost per gallon for kerosene that cost $3.28 for 3.6 gallons.

6. The cost per pound for grapes that cost $1.45 for 1.5 pounds.

7. The cost per pound for a 4.66-pound chicken that sold for $5.75.

8. The cost per pound for a 14-pound turkey that sold for $17.50.

9. The cost per pound for coffee that cost $5.60 for 1.12 pounds.

10. The cost per ounce for perfume that sells at $13.80 for .6 ounce.

Check your answers on page 237.

Solving Multistep Decimal Problems

As you learned in Lesson 19, to solve many problems, you need to perform more than one operation. In this lesson you will work multistep problems involving situations common in everyday life.

Finding Net Pay

Gross pay is the amount of money a person earns before deductions for income taxes, social security (FICA), insurance, union dues, and other items. Net pay, or take-home pay, is the amount a person receives after deductions. To find net pay, first find the sum of all deductions. Then subtract this sum from the gross pay.

Example 1: Using information from the following pay stub, find Anna's net pay.

Employee	SSN	Department	
Anna B. Carr	555-33-5555	Quality Control	
Gross Pay	Federal Income Tax	FICA	Net Pay
$285.00	$52.30	$38.90	$_____

Step 1

$52.30   federal income tax
+ 38.90   FICA
$91.20   total deductions

Step 2

$285.00   gross pay
− 91.20   total deductions
$193.80   net pay

186   UNIT 2: Decimals

STEP 1: Find the sum of Anna's two deductions.

STEP 2: Subtract this sum from Anna's gross pay. Anna's net pay is $193.80.

EXERCISE 34a

Part A. Using information from the following pay stubs, find each person's net pay.

1.

Employee	SSN	Department	
Don E. Fox	123-45-6789	Assembly Line	
Gross Pay	Federal Income Tax	FICA	Net Pay
$392.35	$60.56	$29.81	$_____

2.

Employee	SSN	Department	
Joan K. Lai	303-03-0303	Transportation	
Gross Pay	Federal Income Tax	FICA	Net Pay
$484.02	$77.12	$36.72	$_____

3.

Employee	SSN	Department	
Mark N. Oney	212-21-2122	Engine Maintenance	
Gross Pay	Federal Income Tax	FICA	Net Pay
$512.78	$82.16	$39.22	$_____

4.

Employee	SSN	Department	
Ruben Munoz	987-65-4321	Administration	
Gross Pay	Federal Income Tax	FICA	Net Pay
$840.00	$151.20	$64.26	$_____

5.

Employee	SSN	Department	
Sam Brink	002-00-2200	Personnel	
Gross Pay	Federal Income Tax	FICA	Net Pay
$1460.00	$262.80	$111.69	$_____

Part B. Using the following information, find each person's net pay.

1. Sarah Uhl's gross pay was $410.75. Her deductions were the following: federal income tax, $73.93; FICA, $31.42; insurance, $2.63; and union dues, $3.42. What was Sarah's take-home pay?

2. Vic Jimenez's gross pay was $424.50 last week. His deductions were the following: federal income tax, $76.41; FICA, $32.47; and a pension contribution of $42.90. What was Vic's net pay?

3. Last week Kathy Joos's gross pay was $964.75. The only deductions from her gross pay were the following: federal income tax, $173.66; state income tax, $67.53; and FICA, $73.80. What was Kathy's net pay?

4. Hans Grun's gross pay is $742.60. His last pay stub showed the following deductions: federal income tax, $133.66; state income tax, $51.98; and FICA, $56.80. What was Hans's net pay?

5. Nick Navarro's gross pay for the year amounted to $28,797. His W-2 form showed the following deductions for the year: federal income tax, $5183.46; state income tax, $1439.85; FICA, $2209.71; and health insurance, $780.00. What was Nick's net pay for the year?

Check your answers on page 237.

Making a Budget

A budget shows how a person spends his or her net pay. To make a budget, first find the sum of all expenses, regular and planned. Then subtract this sum from the net pay.

Example 2: Maria's net pay each month is $925.84. She budgets $30 for carfare, $75 for clothes, and $200 for food. Her rent is $436.90. How much does Maria have left?

Step 1		Step 2	
$ 30.00	carfare	$925.84	net pay
75.00	clothes	− 741.90	total expenses
200.00	food	$183.94	left
+ 436.90	rent		
$741.90	total expenses		

STEP 1: Find the sum of Maria's expenses.

STEP 2: Subtract this sum from Maria's net pay. Maria has $183.94 left.

EXERCISE 34b

Answer the questions about the following people's budgets.

1. Leslie's monthly expenses are shown on the right. She takes home $836.52 each month. How much does she have left for other expenses?

Leslie's Monthly Expenses	
Rent	$225.00
Car	94.50
Utilities	45.00
Food	175.00

2. Howard's monthly take-home pay is $1150. His regular expenses include $385.75 for rent, $62.40 for utilities, $225 for food, and a car payment of $136.67. How much does he have left after he pays his regular bills each month?

3. Tim's monthly expenses are shown on the right. Tim takes home $1098 each month. If he puts everthing left over in a savings account, how much does he save each month?

Tim's Monthly Expenses

Rent	$384.50
Utilities	74.00
Food	175.00
Car	124.30
Miscellaneous	125.00

4. Consuelo made a budget to see where her money goes. She takes home $925 per month. Her monthly expenses include rent, $230; utilities, $45; carfare, $50; food, $270; and the babysitter, $325. How much money does Consuelo have left each month after she pays her regular expenses?

5. Terry's take-home pay last year was $14,952. His regular expenses for the year included $4488 for rent, $660 for utilities, $2220 for food, and $2497 for his car. How much did Terry have left after paying his regular expenses?

6. Dan took home $1790.55 last month. His basic expenses included $560 for rent, $86.45 for utilities, and $32.66 for insurance. After paying these expenses, how much money did Dan have for the rest of his expenses?

7. Ingrid and Jake have a combined annual take-home pay of $19,367. Out of this, they pay $6000 in rent, $1020 in utilities, and $322 for apartment insurance. Once these expenses are paid, how much money is left in their budget?

8. Sylvia's monthly expenses include $480 for rent, $82 for utilities, $250 for food, $82 for transportation, and $246 for miscellaneous purchases. If her take-home pay is $2498 each month, how much money does she have left after paying these expenses?

Check your answers on page 238.

Comparing Unit Costs

In Lesson 33 you used the unit-cost formula. If you want to compare two brands or packages to find out which costs less per unit, you can use that formula. First, find the unit cost for each item. If a unit cost does not come out even in the tenths place, include the thousandths place, rounded if necessary. Then, compare the unit costs to see which is lower. (You may want to review how to compare decimals in Lesson 22.)

Example 3:   Which brand of cornflakes costs less per ounce?

   (1) 24 ounces for $1.80
   (2) 20 ounces for $1.40

Step 1	Step 2	Step 3
Unit cost = price ÷ units	Unit cost = price ÷ units	Brand 2 costs less.
Unit cost = $1.80 ÷ 24 oz	Unit cost = $1.40 ÷ 20 oz	
Unit cost = $.075 per oz	Unit cost = $.07 per oz	

STEP 1: Find the cost for the first brand of cornflakes. The unit cost does not come out even in the tenths place, so the answer includes the thousandths place.

STEP 2: Find the unit cost for the second brand of cornflakes. The unit cost comes out even in the tenths place.

STEP 3: Compare the two unit costs and decide which brand costs less.

EXERCISE 34c

Find which item in each pair costs less per unit.

1. Which package of rice costs less per ounce?
   (1) 75 ounces for $2.25
   (2) 60 ounces for $1.50

2. Which container of milk costs less per fluid ounce?
   (1) 120 fluid ounces for $1.80
   (2) 60 fluid ounces for $1.02

3. Which bottle of juice costs less per quart?
   (1) 1.5 quarts for $.90
   (2) 1.6 quarts for $1.00

4. Which package of cereal costs less per ounce?
   (1) 18 ounces for $1.98
   (2) 13 ounces for $1.56

5. Which box of grass seed costs less per pound?
   (1) 4.5 pounds for $3.42
   (2) 3.75 pounds for $3.00

6. Which loaf of bread costs less per pound?
   (1) 1.4 pounds for $1.29
   (2) 2 pounds for $1.49

7. Which piece of cheese costs less per pound?
   (1) .75 pound for $2.56
   (2) .4 pound for $1.44

8. Which turkey costs less per pound?
   (1) 16 pounds for $17.44
   (2) 12 pounds for $11.88

9. Which can of juice costs less per ounce?
   (1) 10 ounces for $1.22
   (2) 8 ounces for $1.04

10. Which can of tuna costs less per ounce?
    (1) 6 ounces for $1.39
    (2) 12 ounces for $2.36

Check your answers on page 238.

Solving Perimeter, Circumference, and Area Multistep Problems

Often you need to find a perimeter, circumference, or area in order to solve another problem. Two examples of such problems follow.

Example 4: Gillian is planning to fence her backyard, which measures 10 yards by 20 yards. One yard of fencing costs $22.50. How much will she pay for the fence?

Step 1	**Step 2**
10 yards	Cost = number × rate
20 yards	Cost = 60 yards × $22.50
10 yards	Cost = $1350
+ 20 yards	
60 yards	

STEP 1: Find the perimeter of the the back yard. That lets you know how much fencing Gillian needs.

STEP 2: Use the cost formula to find the total cost of the fencing. Substitute *60 yards* for *number* and *$22.50* for *rate*. Gillian will pay $1350 for 60 yards of fencing.

Example 5: One gallon of paint covers 150 square meters. How many gallons of paint are needed to cover 4 walls that each measure 15.75 meters by 4 meters?

Step 1	**Step 2**	**Step 3**
Area = length × width	63 sq m × 4 = 252 sq m	252 sq m ÷ 150 = 1.68 gal
Area = 15.75 m × 4 m		
Area = 63 sq m		

STEP 1: Use the formula for the area of a rectangle to find the area of one wall.

STEP 2: Multiply the area of one wall by 4 to find the total area to be painted.

STEP 3: Divide the total area by the area covered by 1 gallon. The result is the number of gallons it will take to cover the 4 walls.

EXERCISE 34d

Use your math skills to solve each of the following problems.

1. A circular track has a diameter of 200 yards. How many yards does Janna run when she goes around the track 8 times?

2. It costs $5.56 per square meter to resurface a wooden floor. How much will it cost to resurface a floor that measures 8 meters by 5.5 meters?

3. Kevin must fence his rectangular cornfield that measures .5 km by .75 km. Fencing costs $5000 per kilometer. How much will it cost him to fence the field?

4. A pool cover is on sale for $.45 per square foot. How much will it cost to cover the surface of a rectangular pool that measures 25 feet by 15 feet?

5. A tarpaulin used to cover a baseball field measures 40 yards by 40 yards. What is the total weight of the tarpaulin if it weighs 2.8 pounds per square yard?

6. Raymond is painting a fence that measures 30 yards by 5 yards. If a gallon of paint covers 20 square yards, how much paint will he use on the fence?

7. Martha is carpeting a room. The dimensions of the room are shown in the diagram. The carpet costs $23.50 per square yard. How much will it cost to carpet the room?

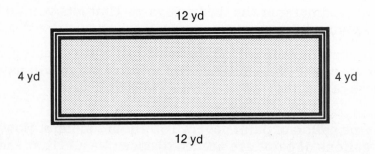

12 yd

4 yd

4 yd

12 yd

8. Artie built a basketball court that measured 75 feet by 32 feet. The court cost $3.20 per square foot. What was the total cost for the court?

Check your answers on page 238.

DECIMALS REVIEW

Part A. Solve each problem.

1. $53 + 2.6 =$
2. $62.48 - 31.4 =$
3. $.42 \times .03 =$
4. $.81 \div 9 =$
5. $.5651 - .5 =$
6. $.84 - .065 =$
7. $20 \times .8 =$
8. $8.4 + .011 =$
9. $.096 \div .3 =$
10. $457.5 \div 5 =$
11. $.18 \times .6 =$
12. $7.9 + 40 =$
13. $.0039 + 9.3 =$
14. $.84 - .065 =$
15. $.39 \times .59 =$
16. $23.68 \div 3.2 =$
17. $.93 - .35 =$
18. $7.135 \times 9.2 =$
19. $.1742 \div .66 =$
20. $6.06 + 26.2 =$
21. $42.285 - 1.991 =$
22. $.4578 \div 327 =$
23. $2405 \times 30.1 =$
24. $2.68 + 59.3 =$
25. $63.08 - 2.409 =$
26. $.435 + .65 =$
27. $40,041 - 8.252 =$
28. $623.4 \times 5.03 =$
29. $17.5 \div 100 =$
30. $712.8 + .76 + 40.2 =$
31. $874.644 - 22.9381 =$
32. $3000 \times .72 =$
33. $208.05 \div .625 =$

Part B. Solve each problem.

1. Write each of the following in numbers:
   (a) six hundredths
   (b) twenty-four thousandths
   (c) four and twenty-five hundredths
   (d) fifty-five thousandths
   (e) thirty and three tenths

2. Write these decimals in order from smallest to largest:
   4.54, 4.35, 4.34, 4.53

3. Write these decimals in order from largest to smallest:
   7.28   7.99   7.42   7.038

4. Round these decimals as directed:
   (a) 32.464 to the nearest hundredth
   (b) .3942 to the nearest tenth
   (c) 73.004 to the nearest hundredth
   (d) $82.156 to the nearest cent
   (e) .55 to the nearest tenth

5. How much does 2.25 pounds of meat cost at $3.49 per pound?

6. What is the unit cost for candy that sells for $1.99 for 8 ounces?

7. Find the average of the following numbers: 8.5, .98, 84, 7.5, and 1.93.

8. What is the perimeter of a square whose sides are each 1.2 centimeters long?

9. What is the area of a rectangle that is 5.5 kilometers by 6.5 kilometers?

10. Find the perimeter and the area of this triangle.

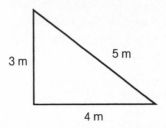

11. Find the circumference and the area of a circle with a radius of 3.1 centimeters.

12. Jerry's gross pay was $410.75. His deductions were the following: federal income tax, $73.91; FICA, $31.41; insurance, $5.63; and union dues, $4.44. What was Jerry's take-home pay?

13. Ali's monthly take-home pay is $1150. His regular expenses include $385.75 for rent, $62.40 for utilities, and $225 for food. How much does he have left after he pays his regular bills each month?

14. Which brand of cornflakes costs less per ounce?
    (1) 24 ounces for $1.80
    (2) 20 ounces for $1.40

15. A circular track has a radius of 20 yards. How many yards does Herb run when he goes around the track twice?

Check your answers on page 239.

GED PRACTICE 2

This section will give you practice in answering questions like those on the GED. The 20 questions in this Practice are all multiple-choice, like the ones on the GED. As you do this Practice, use the skills you've studied in this unit.

Directions: Choose the <u>one best answer</u> to each item.

1. Mr. Chen collected the following rents from his tenants: $265.50, $434, $376.50, and $282. What was the total amount he collected?
   (1) $1358.50
   (2) $1358
   (3) $1357
   (4) $1147.50
   (5) $ 339.50

2. Jeff bought a pair of trousers on sale for $62.50. The original price had been $189. How much money did Jeff save by buying on sale?
   (1) $127.50
   (2) $127.00
   (3) $126.50
   (4) $125.75
   (5) $ 60.61

3. What is the thickness (in centimeters) of 15 sheets of paper if one sheet is .023 centimeters thick?
   (1)   .0335
   (2)   .0345
   (3)   .128
   (4)   .345
   (5) 15.023

4. Three friends shared the cost of a $44.59 present for George. How much did each person pay?
   (1) $ 11.15
   (2) $ 14.86
   (3) $ 14.87
   (4) $ 44.59
   (5) $133.77

5. How many gallons of gasoline can you buy for $16.98 if one gallon costs $.849?
   (1)   .02
   (2)   .2
   (3)   2
   (4)   20
   (5) 200

6. Jake earns $6.35 per hour at a part-time job. One week he worked the following hours: Monday, 4 hours; Tuesday, 3 hours; Wednesday, 2 hours; Thursday, 6 hours; and Friday, 4 hours. How much money did Jake earn that week?
   (1) $ 24.13
   (2) $120.65
   (3) $127.00
   (4) $130.00
   (5) $500.00

7. Which of the following is equal to 2.2 liters?
   (1) .0022 kL
   (2) .0022 daL
   (3) .0022 dL
   (4) .0022 cL
   (5) .0022 mL

8. The rainfall during one week was as follows: Sunday, .1 inch; Monday, 1.1 inches; Tuesday, 2 inches; Wednesday, .9 inch; Thursday, .6 inch; Friday, .2 inch; and Saturday, .9 inch. What was the average rainfall (in inches) per day that week?
   (1)   .828
   (2)   .829
   (3) 5.714
   (4) 5.8
   (5) 40

9. A rectangular garden has sides that measure 20.5 meters, 30 meters, 20.5 meters, and 30 meters. How many meters of fencing are needed to enclose the garden?

(1) 100
(2) 101
(3) 307.5
(4) 470
(5) 615

10. What is the perimeter (in centimeters) of a triangle whose sides are each 2.8 centimeters long?

(1) 3.92
(2) 7.84
(3) 8.4
(4) 11.2
(5) 84

11. How many square meters of tile are needed to cover a courtyard floor that measures 12.5 meters by 8.2 meters?

(1) 10.25
(2) 20.7
(3) 41.4
(4) 51.25
(5) 102.5

12. Find the circumference (in centimeters) of a circle whose radius is 6.5 centimeters.

(1) 10.205
(2) 20.41
(3) 40.82
(4) 132.665
(5) 530.66

13. What is the area (in square feet) of a circular tabletop whose radius is 3.5 feet?

(1) 10.99
(2) 21.98
(3) 34.509
(4) 38.465
(5) 153.86

14. How much does 7 pounds of apples cost at .79 per pound?

(1) $ .79
(2) $4.93
(3) $4.96
(4) $5.46
(5) $5.53

15. What is the cost per ounce for cereal that sells at $1.20 for a 15-ounce box?

(1) $ .08
(2) $ .80
(3) $ 8.00
(4) $ 18.00
(5) $180.00

16. Antonio's gross pay last week was $632.90. His deductions were the following: $126.58 for federal taxes; $47.47 for FICA; $31.60 for state taxes; and $11.75 for union dues. What was Antonio's net pay?

(1) $217.40
(2) $415.50
(3) $527.70
(4) $537.70
(5) $850.30

17. Juan's take-home pay is $1000 a month. His regular expenses are the following: rent, $423.75; utilities, $52.30; and transportation, $46. After he takes care of his regular expenses, how much does he have left for food and other expenses each month?

(1) $ 174.02
(2) $ 476.05
(3) $ 477.95
(4) $ 522.05
(5) $1522.05

18. Which of the following containers of bread crumbs is least expensive per ounce?

(1) 15 ounces for $1.60
(2) 16 ounces for $1.65
(3) 20 ounces for $2.05
(4) 24 ounces for $2.50
(5) 36 ounces for $4.00

19. A one-way ticket costs $1.55. A book of 20 one-way tickets costs $28.75. How much money can you save per one-way trip if you buy the 20-ticket book?

(1) $ .01
(2) $ .11
(3) $ 1.10
(4) $ 1.33
(5) $27.20

20. Joaquin is painting a wall that measures 10 meters by 2 meters. If a liter of paint covers 10 square meters, how many liters of paint will he use on the wall?

   (1) 1.2
   (2) 2
   (3) 2.4
   (4) 12
   (5) 20

Check your answers on page 239.

GED PRACTICE 2 SKILLS CHART

To review the mathematics skills covered by the items in GED Practice 2, study the following lessons in Unit 2.

Unit 2	Decimals	Item Number
Lesson 24	Adding Decimals and Mixed Decimals	1
Lesson 25	Subtracting Decimals and Mixed Decimals	2
Lesson 26	Multiplying Decimals and Mixed Decimals	3
Lesson 27	Dividing Decimals and Mixed Decimals	4, 5
Lesson 28	Solving Decimal Word Problems	6
Lesson 29	Metric Measurements	7
Lesson 30	Finding Averages	8
Lesson 31	Finding Perimeter and Area with Metric Measurements	9, 10, 11
Lesson 32	Finding the Circumference and Area of Circles	12, 13
Lesson 33	Using the Cost Formula	14, 15
Lesson 34	Solving Multistep Decimal Problems	16, 17, 18, 19, 20

Posttest

This posttest will help you review the work you've done. It will also give you practice in answering questions like those on the GED. It is made up of 28 multiple-choice questions, half as many questions as there are on the Mathematics Test of the GED. The skills chart on page 204 will help you plan a review of the skills covered by this book that you may need to practice further.

Directions: Choose the one best answer to each item.

1. In each of 6 practices, Joan ran 5039 feet. How many miles did she run altogether?

   (1)   .573
   (2)   .95
   (3)   5.73
   (4)   9.54
   (5)   57.3

2. One carton of paper towels weighs 1 pound 9 ounces. How much does a case of 24 cartons weigh?

   (1) 13 pounds 8 ounces
   (2) 24 pounds 9 ounces
   (3) 25 pounds
   (4) 37 pounds 8 ounces
   (5) 240 pounds

3. Susan sleeps 7 hours 45 minutes each night. How much does she sleep in a week?

   (1) 49 hours
   (2) 49 hours 45 minutes
   (3) 49 hours 325 minutes
   (4) 54 hours 15 minutes
   (5) 54 hours 25 minutes

4. A truck is carrying 7 identical cars that weigh 6 tons 250 pounds altogether. How much does each car weigh?

   (1) 892 pounds
   (2) 1750 pounds
   (3) 1 ton 35 pounds
   (4) 2 tons
   (5) 42 tons 1750 pounds

5. A triangular lot measures 75 meters by 100 meters by 125 meters. What is its perimeter (in meters)?

   (1)   150
   (2)   300
   (3)   468.75
   (4)   7,625
   (5) 937,500

Item 6 is based on the following figure.

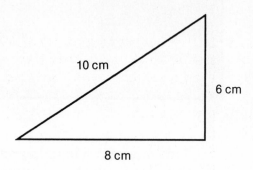

6. What is the area of the triangle?

   (1) 24 centimeters
   (2) 24 square centimeters
   (3) 40 square centimeters
   (4) 48 square centimeters
   (5) 80 square centimeters

Items 7 and 8 are based on the following sketch of a lot with a garage.

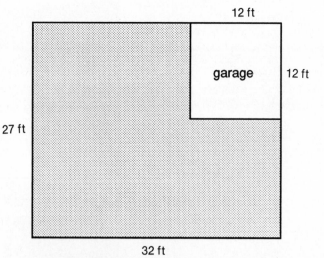

7. How many feet of fencing are needed to fence around the shaded portion of the lot? (Note: No fencing is needed by the walls of the garage.)

   (1)   24
   (2)   77
   (3)   94
   (4) 118
   (5) 864

8. What is the area (in square feet) of the shaded portion of the lot?

   (1)   77
   (2)   94
   (3) 118
   (4) 720
   (5) 864

9. Six volunteers sold a total of 2454 tickets to a benefit concert. What was the average number of tickets sold by each volunteer?

   (1)   49
   (2)   309
   (3)   409
   (4)   2,460
   (5)   14,724

10. Clare's computer screen can show 80 characters per line and 23 lines. How many characters can the screen display at once?

    (1) 1600
    (2) 1760
    (3) 1840
    (4) 2400
    (5) 2560

11. In one year, George made 287 trips on his rural mail route. If each trip was 34 miles, how many miles did he drive that year?

    (1)   321
    (2)   2,009
    (3)   9,758
    (4)   9,938
    (5)   87,548

12. An average of 30,460 fans attended each of 18 Pittsburgh Pirates games last season. How many fans attended all season?

    (1)   54,920
    (2)   54,880
    (3)   274,140
    (4)   541,400
    (5)   548,280

13. Mario must provide a safety training class for 2784 employees at the factory where he works. If the training room seats 32, how many class sessions must he hold?

    (1)   86 r1
    (2)   87
    (3) 802 r20
    (4) 807
    (5) 861

14. Of the 9000 people who attended opening day of the golf tournament, 783 bought student tickets. The rest were regular admissions. How many regular admissions were there?

    (1) 8106
    (2) 8107
    (3) 8217
    (4) 9017
    (5) 9106

15. Smithberg has 6780 families. According to the census, 7973 boys and 8638 girls live in the town. What is the average number of children per household?

    (1) 1.18
    (2) 1.27
    (3) 2
    (4) 2.45
    (5) 3

16. The Kleins are moving into a new house that is 3 times the size of their first apartment. Their new house has 750 square feet on each of 2 floors. How large was their first apartment?

    (1)   150 square feet
    (2)   250 square feet
    (3)   500 square feet
    (4) 1500 square feet
    (5) 2250 square feet

17. A famous rock star was paid $15,000,000 to appear in 3 television commercials for a soft-drink company. Each commercial required 8 hours of her time. How much did she make per hour?

    (1) $ 487,500
    (2) $ 600,000
    (3) $ 625,000
    (4) $1,875,000
    (5) $5,000,000

18. At the end of a trading day, the prices of these 5 stocks changed in the following ways:

Stock A   +.25
Stock B   +.50
Stock C   +.125
Stock D   +.375
Stock E   +.75

Which of the following correctly lists the changes from smallest to largest?

(1) A, B, C, D, E
(2) B, C, A, E, D
(3) C, A, D, B, E
(4) E, B, D, A, C
(5) E, B, D, C, A

19. Jose runs for .75 hour at 12 kilometers per hour 5 days each week. How many meters does he run in a week?

(1)   4.5
(2)   45
(3)   450
(4)   4,500
(5)   45,000

20. The vehicles owned by the neighbors on Kelly St. get the following miles per gallon:

Hernadez' jeep	15.6 mpg
Noonans' sedan	22.4 mpg
Blitskys' RV	8.1 mpg
Porters' motorcycle	39.8 mpg
Hawks' pickup truck	19.3 mpg

What is the average miles per gallon (mpg) of *all* the vehicles?

(1)   15.65
(2)   16.35
(3)   21.04
(4)   21.4
(5) 105.2

21. Harold rode his lawn mower around the edge of a square lot that measures 34 meters on each side. How many meters did Harold ride?

(1)   51
(2)   102
(3)   106.76
(4)   136
(5) 1156

Items 22 and 23 are based on this sketch of a fenced swimming pool.

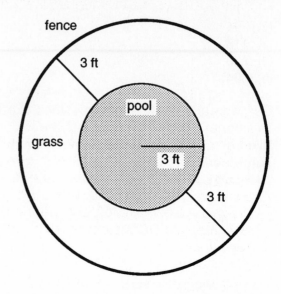

22. To make a cover for the pool, how many pieces of plywood 6 feet by 8 feet are needed?

(1) 1
(2) 2
(3) 3
(4) 4
(5) 32

23. Rounded to the nearest foot, how many feet of fencing will it take to enclose the pool with a circular fence that is 3 feet from the edge of the pool?

(1)   19
(2)   29
(3)   38
(4)   64
(5) 114

24. Margaret wants to cover a round canister lid with vinyl wallpaper to match her walls. The canister has a radius of 6 centimeters. How many square meters of vinyl does she need?

(1)   .0942
(2) 1.13
(3) 9.42
(4) 18.84
(5) 37.68

25. One pound of lobster costs $12.85. How much change would you get from a $20 bill if you used it to pay for a 1.4-pound lobster?

    (1) $ 2.01
    (2) $ 5.75
    (3) $ 7.15
    (4) $17.99
    (5) $28.00

26. Michael's net pay is $246.35 per week. His monthly bills include rent at $475 per month, utilities at $60.25 per month, and transportation at $48.80 per month. How much does he have left over after his bills are paid? (Consider that 1 month = 4 weeks.)

    (1) $337.70
    (2) $401.35
    (3) $402
    (4) $584.05
    (5) $985.40

27. Which of the following costs least per ounce?

    (1) 6 ounces for $ .87
    (2) 12 ounces for $1.70
    (3) 15 ounces for $2.24
    (4) 24 ounces for $3.29
    (5) 24 ounces for $3.50

Check your answers on page 240.

Item 28 is based on the following sketch of an L-shaped room.

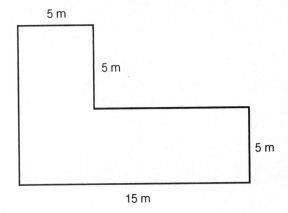

28. At $25 per square meter, how much would it cost to carpet the room?

    (1) $ 50
    (2) $ 100
    (3) $ 875
    (4) $1000
    (5) $2500

MATHEMATICS POSTTEST SKILLS CHART

To review the mathematics skills covered by the items in the Posttest, study at least the following lessons.

Unit 1	Whole Numbers	Item Number
Lesson 13	Units of Measure	1
Lesson 14	Working with Units of Measure: The English System	2, 3, 4
Lesson 15	Finding Perimeter and Area	5, 6
Lesson 16	Working with Perimeter and Area	7, 8
Lesson 17	Finding Averages	9
Lesson 18	Solving Whole-Number Word Problems	10, 11, 12, 13
Lesson 19	Solving Multistep Whole-Number Word Problems	14, 15, 16, 17

Unit 2	Decimals	Item Number
Lesson 28	Solving Decimal Word Problems	18
Lesson 29	Metric Measurements	19
Lesson 30	Finding Averages	20
Lesson 31	Finding Perimeter and Area with Metric Measurements	21
Lesson 32	Finding the Circumference and Area of Circles	22, 23, 24
Lesson 33	Using the Cost Formula	25
Lesson 34	Solving Multistep Decimal Problems	26, 27, 28

Answers and Solutions

In this section are the answers for all the problems in this book. To help you check your work, the solutions for many of the problems are shown.

PRETEST (page 1)

1. The value of the 5 in 4750 is **50**, or **5 tens.**

2. **fourteen thousand, sixty-five**

3. **106,075**

4. **11, 100, 101, 111, 1001, 1010, 1011, 1110**

5. **12,100**   6. **9**

7. **37**   8. **378**

9. **89**   10. **622**

11. **1397**   12. **7449**

13. **2**   14. **3**

15. **24**   16. **1531**

17. **1128**   18. **7 × 6 = 42**

19. **20**   20. **86**

21. **929,740**   22. **71,370**

23. **3**   24. **2**

25. **51**   26. **2030**

27. (a) **2 pt**   (d) **1 mi**
    (b) **2 lb**   (e) **365 da**
    (c) **3 ft**   (f) **2000 lb**

28. **(3)**

29. Jorge can cut **3 shelves** from the board.
    Jorge's board is 96 in. long:
       8 ft × 12 = 96 in.
    It will make three 32-in. shelves:
       96 in. ÷ 32 in. = 3.

30. (a) Kurt needs **140 ft** of fencing.
       40 ft + 30 ft + 40 ft + 30 ft = 140 ft
    (b) **Yes**, the fence will allow enough room.
       30 ft × 40 ft = 1200 sq ft, which is much more than the vet says the puppy needs.

31. The average high temperature that week was **89**.
    87 + 89 + 91 + 88 + 86 + 92 + 90 = 623;
    623 ÷ 7 = 89

32. It will cost **$9** for 24 oz of shrimp.
    1 lb = 16 oz, so $6 ÷ 16 oz = $.375 per oz;
    $.375 × 24 oz = $9

33. The **8** is in the thousandths place.

34. **Fifty-two hundredths**

35. **.0011, .11, 1.001, 1.01, 1.0101, 1.11, 10.11, 11.1**

36. **2.9**

37. **$9.25**   38. **9589.2**

39. **989.15**   40. **11.7**

41. **3.488**   42. **1.1**

43. **34**   44. **509,218**

45. Sandra will make **$192.50** for the week.
    6 hr + 7 hr + 7 hr + 8 hr + 7 hr = 35 hr;
    35 hr × $5.50 = $192.50

46. 1 g of salt is the same as **1000 mg.**

47. The average daily rainfall during that period was **1.022 in.**
    .87 in. + .5 in. + .73 in. + 1.01 in. + 2 in. = 5.11 in.
    5.11 in. ÷ 5 = 1.022 in.

48. The area of the triangle is **9.8 sq m.**

    $$\text{Area} = \frac{\text{base} \times \text{height}}{2}$$

    $$\text{Area} = \frac{7 \text{ m} \times 2.8 \text{ m}}{2}$$

    $$\text{Area} = 9.8 \text{ sq m}$$

49. The circumference of the circle is **13.8 cm.**

    Diameter = 4.4 cm

    Circumference = 3.14 × diameter

    Circumference = 3.14 × 4.4 cm

    Circumference = 13.816 cm, or 13.8 cm rounded to the nearest tenth

  The area of the circle is **15.2 sq cm.**

    Area = 3.14 × radius × radius

    Area = 3.14 × 2.2 cm × 2.2 cm

    Area = 15.1976 sq cm, or 15.2 sq cm rounded to the nearest tenth

50. Can **(2)** costs less per ounce.

  (1) $.89 ÷ 6 oz = $.148 per ounce

  (2) $1.75 ÷ 12 oz = $.145 per ounce

UNIT 1   WHOLE NUMBERS

EXERCISE 1a (PAGE 9)

Part A

1. **22   56**
2. **523   142   987   900**
3. (a) **12**   (b) **3**   (c) **402**   (d) **89**
   (e) **98**   (f) **6**
4. (a) **12**   (b) **67**   (c) **312**   (d) **831**
   (e) **509**   (f) **33**
5. (a) **505**   (b) **678**   (c) **943**   (d) **142**
   (e) **200**   (f) **607**
6. (a) **5 tens**, or **50** (b) **6 ones**, or **6**
   (c) **3 hundreds**, or **300** (d) **1 one**, or **1**
   (e) **5 ones**, or **5** (f) **7 tens**, or **70**

Part B

1. 42 has **4** tens and **2** ones.
2. 31 has **3** tens and **1** one.
3. 93 has **9** tens and **3** ones.
4. 80 has **8** tens and **0** ones.
5. 55 has **5** tens and **5** ones.
6. 67 has **6** tens and **7** ones.
7. 689 has **6** hundreds, **8** tens, and **9** ones.
8. 401 has **4** hundreds, **0** tens, and **1** one.
9. 513 has **5** hundreds, **1** ten, and **3** ones.
10. 400 has **4** hundreds, **0** tens, and **0** ones.
11. 999 has **9** hundreds, **9** tens, and **9** ones.
12. 35 has **0** hundreds, **3** tens, and **5** ones.

Part C

1. The 8 is in the **ones** place, so it has a value of **8**.
2. The 1 is in the **tens** place, so it has a value of **10**.
3. The 2 is in the **hundreds** place, so it has a value of **200**.

Part D

1. The 6 is in the **ones** place, so it has a value of **6**.
2. The 0 is in the **tens** place, and it has a value of **0**.
3. The 7 is in the **hundreds** place, so it has a value of **700**.

Word Problem

  It will take **4** ten-dollar bills and **2** one-dollar bills to buy the sweater.

EXERCISE 1b (PAGE 11)

Part A

1. (a) **46,143**   (b) **5728**   (c) **56,340**
   (d) **7000**   (e) **19,037**   (f) **3406**
2. (a) **42,765**   (b) **89,001**   (c) **21,277**
   (d) **90,000**   (e) **76,142**   (f) **98,642**
3. (a) **5 hundreds**, or **500**
   (b) **1 thousand**, or **1000**
   (c) **1 hundred thousand**, or **100,000**
   (d) **8 ten thousands**, or **80,000**
   (e) **3 ten millions**, or **30,000,000**
   (f) **7 billions**, or **7,000,000,000**

Part B

	billions	hundred millions	ten millions	millions	hundred thousands	ten thousands	thousands	hundreds	tens	ones
1.						4	2	7	9	3
2.						9	6	2	0	0
3.							6	1	8	5
4.						3	1	8	2	7
5.							1	4	5	9
6.		1	6	9	9	6	3	0	0	0
7.					8	5	4	0	4	6
8.							4	1	4	1
9.	2	0	0	9	6	8	3	0	0	0
10.		9	3	0	0	0	0	0	0	

Part C

1. 1536 has **1** thousand, **5** hundreds, **3** tens, and **6** ones.

2. 32,709 has **3** ten thousands, **2** thousands, **7** hundreds, **0** tens, and **9** ones.

3. 347,296 has **3** hundred thousands, **4** ten thousands, **7** thousands, **2** hundreds, **9** tens, and **6** ones.

4. 9,863,074 has **9** millions, **8** hundred thousands, **6** ten thousands, **3** thousands, **0** hundreds, **7** tens, and **4** ones.

5. 20,968,147 has **2** ten millions, **0** millions, **9** hundred thousands, **6** ten thousands, **8** thousands, **1** hundred, **4** tens, and **7** ones.

6. 3,298,645,009 has **3** billions, **2** hundred millions, **9** ten millions, **8** millions, **6** hundred thousands, **4** ten thousands, **5** thousands, **0** hundreds, **0** tens, and **9** ones.

Part D

1. The 4 is in the **ones** place, so it has a value of **4**.

2. The 0 is in the **tens** place, and it has a value of **0**.

3. The 8 is in the **hundreds** place, so it has a value of **800**.

4. The 6 is in the **thousands** place, so it has a value of **6000**.

Part E

1. The 1 is in the **ones** place, so it has a value of **1**.

2. The 4 is in the **tens** place, so it has a value of **40**.

3. The 7 is in the **hundreds** place, so it has a value of **700**.

4. The 5 is in the **thousands** place, so it has a value of **5000**.

5. The 6 is in the **ten-thousands** place, so it has a value of **60,000**.

6. The 3 is in the **hundred-thousands** place, so it has a value of **300,000**.

7. The 2 is in the **millions** place, so it has a value of **2,000,000**.

8. The 0 is in the **ten-millions** place, and it has a value of **0**.

9. The 9 is in the **hundred-millions** place, so it has a value of **900,000,000**.

10. The 8 is in the **billions** place, so it has a value of **8,000,000,000**.

Word Problem

Your place-value chart should look something like the following:

ten thousands	thousands	hundreds	tens	ones
1	**9**	**0**	**3**	**2**

EXERCISE 2 (PAGE 15)

Part A

1. (2)	2. (2)	3. (3)
4. (4)	5. (1)	6. (2)

Part B

1. **3,418,623**	2. **648,000**	3. **5,200,000**
4. **3010**	5. **10,449**	6. **33,492**
7. **16,856**	8. **41,937**	9. **85,020**
10. **22,221**		

Part C

1. **five thousand, two hundred forty**
2. **eight thousand, forty-four**
3. **eleven thousand, one**
4. **twenty-six thousand, seven hundred thirty-two**
5. **two hundred ninety-five thousand, six hundred thirty-one**
6. **three hundred twenty-four thousand, seven hundred sixty-eight**
7. **six million, six hundred sixty-one thousand, six hundred thirty-one**
8. **two hundred twenty-one million, three hundred twenty thousand**
9. **forty billion, seven million, two hundred thirteen**
10. **three billion, four hundred two million, three hundred thirty-three thousand, nine hundred ninety-eight.**

Word Problem

There were **21,973** male officers and **2087** female officers.

EXERCISE 3 (PAGE 17)

Part A

1. **8, 56, 147, 238, 433, 4004**
2. **60, 82, 555, 915, 6921, 9000**
3. **2, 90, 357, 537, 2004, 7002**
4. **102, 201, 466, 500, 2001, 4006**

Part B

1. 342, 324, 314, 93, 71, 52, 43
2. 7034, 861, 734, 304, 201, 128, 9
3. 3894, 984, 894, 808, 666, 99, 66
4. 4060, 4006, 820, 406, 146, 76, 46

Word Problem

Arnie would list the fish in this order:
black sea bass, Spanish mackerel, giant sea bass, white shark.

EXERCISE 4 (PAGE 20)

Part A

1. (1)   2. (1)   3. (2)   4. (1)   5. (2)
6. (1)   7. (2)
8. (2)   9. (1)
10. (2)

Part B

1. 700   2. $500   3. 3000   4. 7000
5. 92,000   6. 81,770   7. 500,000
8. 654,400   9. 8,000,000   10. 5600
11. 20,000   12. 45,000

Part C

1. 5,000,000   2. 4,700,000   3. 4,720,000

Word Problem

Peter wrote **3000** nautical miles in his notebook. Rounded to the nearest thousand, 3069 is closer to 3000 than to 4000.

EXERCISE 5a (PAGE 25)

1. 4   2. 9   3. 8   4. 6
5. 8   6. 8   7. 7   8. 7
9. 9   10. 6   11. 7   12. 6
13. 3   14. 9   15. 9   16. 8
17. 9   18. 2   19. 5   20. 9

Word Problem

Margie scored **5 goals** in all that week.
(2 + 3 = 5)

EXERCISE 5b (PAGE 26)

1. 85   2. 99   3. 79   4. 87
5. 77   6. 89   7. 67   8. 49
9. 34   10. 89   11. 38   12. 78
13. 86 14. 99   15. 97   16. 76
17. 88 18. 84   19. 72   20. 88

Word Problem

Fred worked **49 hours** in all.
(38 + 11 = 49)

EXERCISE 5c (PAGE 27)

1. 27   2. $39   3. 25   4. 59
5. $48   6. 69   7. 59   8. $99
9. 88   10. 95   11. $58   12. 39
13. $17 14. 29   15. 19   16. 36
17. 79   18. 95   19. $19   20. 68

Word Problem

Juan and Maria have **78 CDs** altogether.
(46 + 32 = 78)

EXERCISE 5d (PAGE 28)

1. 978   2. 385   3. 958
4. 897   5. $777   6. $699
7. 238   8. 859   9. $703
10. 469 11. 757   12. 929
13. 958 14. 999   15. 883
16. 789 17. 591   18. 552
19. 780 20. 878

Word Problem

The police ticketed **458 cars** in all.
(342 + 116 = 450)

EXERCISE 5e (PAGE 29)

1. 8   2. 9   3. 8
4. 7   5. 9   6. 7
7. 9   8. 9   9. 8
10. 7   11. 9   12. 9

Word Problem

Gerry got a total of **6 hits** that week.
(1 + 3 + 2 + 0 = 6)

EXERCISE 5f (PAGE 30)

1. 58   2. $78   3. 98
4. 63   5. 88   6. 39
7. 63   8. 94   9. 69
10. 38 11. 66   12. 99
13. 99 14. 68   15. 89
16. 66 17. 36   18. 58
19. 97 20. 99

Word Problem

Mr. Fiedler sold **89 hot dogs** in all that Tuesday. (11 + 35 + 43 = 89)

EXERCISE 5g (PAGE 31)

1. **8569**   2. **$9164**
3. **6588**   4. **8988**
5. **9799**   6. **5559**
7. **7489**   8. **6976**
9. **7489**   10. **8333**
11. **8735**   12. **9756**
13. **6998**   14. **4398**
15. **$6649** 16. **8686**

Word Problem

Beverly paid a total of **$769** for rent, food, and utilities that month.
($416 + $242 + $111 = $769)

EXERCISE 6a (PAGE 33)

1. **24**   2. **32**   3. **$42**   4. **23**
5. **91**   6. **23**   7. **91**   8. **74**
9. **30**   10. **21**   11. **41**   12. **$11**
13. **51**   14. **31**   15. **62**   16. **47**
17. **52**   18. **72**   19. **80**   20. **$68**

Word Problem

A total of **25 inches** of snow fell in January and February. (17 + 8 = 25)

EXERCISE 6b (PAGE 34)

1. **$116**   2. **129**   3. **388**
4. **228**   5. **859**   6. **1359**
7. **1328**   8. **$1510** 9. **326**
10. **960**   11. **2378** 12. **$1666**
13. **4046** 14. **5143** 15. **4489**
16. **$3524** 17. **6489** 18. **8077**
19. **9044** 20. **9677**

Word Problem

There is **$1339** in their accounts combined. ($715 + $624 = $1339)

EXERCISE 6c (PAGE 35)

1. **921**   2. **1242**   3. **$1223**
4. **1332**   5. **6428**   6. **$6571**
7. **10,233** 8. **6933**
9. **10,876** 10. **11,196**
11. **1389**   12. **1544**
13. **$73,855** 14. **66,221**
15. **13,994** 16. **68,752**

Word Problem

Manny will have earned **$11,802**.
($2987 + $3892 + $4923 = $11,802)

EXERCISE 7a (PAGE 39)

1. **1**   2. **2**   3. **4**   4. **$6**
5. **2**   6. **$1**   7. **5**   8. **3**
9. **2**   10. **6**   11. **5**   12. **4**
13. **$9** 14. **1**   15. **3**   16. **1**
17. **8**   18. **1**   19. **6**   20. **3**

Word Problem

Annette spent **$2** less on Tuesday.
($8 − $6 = $2)

EXERCISE 7b (PAGE 40)

1. **51**   2. **$11**   3. **26**   4. **34**
5. **34**   6. **16**   7. **51**   8. **$21**
9. **25**   10. **43**   11. **11**   12. **22**
13. **$20** 14. **25**   15. **1**   16. **19**
17. **61**   18. **32**   19. **$53** 20. **52**

Word Problem

Jack had **$14** left. ($28 − $14 = $14)

EXERCISE 7c (PAGE 41)

1. **1041**   2. **$1411** 3. **5416**
4. **3210**   5. **1000**   6. **2222**
7. **3201**   8. **1112**   9. **$1111**
10. **2107** 11. **1242** 12. **1310**
13. **$5431** 14. **1513** 15. **7350**
16. **2650** 17. **2212** 18. **7517**
19. **6202** 20. **$1012**

Word Problem

In Raphael's collection, **169 baseball cards** remained. (1569 − 1400 = 169)

EXERCISE 8a (PAGE 43)

1. **$9**   2. **7**   3. **9**
4. **8**   5. **8**   6. **6**
7. **$9**   8. **9**   9. **9**
10. **7**   11. **24** 12. **28**
13. **19**   14. **$27** 15. **8**
16. **17**   17. **$29** 18. **16**
19. **39**   20. **$56**

Word Problem

Felix had **$12** left. ($21 − $9 = $12)

EXERCISE 8b (PAGE 44)

1. **522**   2. **$441** 3. **252**
4. **159**   5. **273**   6. **99**
7. **155**   8. **112**   9. **281**
10. **798** 11. **2296** 12. **4111**

13. **5291**	14. **5273**	15. **1181**
16. **5941**	17. **4612**	18. **1325**
19. **4750**	20. **122**	

Word Problem

Do-Rite employs **2123 more people** this year than last. (7086 − 4963 = 2123)

EXERCISE 8c (PAGE 45)

1. **367**	2. **595**	3. **158**
4. **201**	5. **2064**	6. **1151**
7. **1669**	8. **1947**	9. **12,457**
10. **2157**	11. **28,821**	12. **4989**
13. **21,869**	14. **354,568**	
15. **397,189**	16. **249,460**	

Word Problem

The restaurant served **215 fewer meals** this week. (2000 − 1785 = 215)

EXERCISE 8d (PAGE 47)

1. **299**	2. **152**	3. **29**
4. **298**	5. **189**	6. **2179**
7. **4722**	8. **3448**	9. **2991**
10. **5889**	11. **1189**	12. **3093**
13. **6892**	14. **7916**	15. **1199**
16. **23,929**	17. **20,158**	18. **16,089**
19. **79,017**	20. **10,309**	

Word Problem

The price of the car had decreased **$1762**. ($15,246 − $13,484 = $1762)

MIXED PRACTICE 1 (page 48)

1. **7**	2. **54**	3. **7**
4. **31**	5. **32**	6. **15**
7. **9**	8. **10**	9. **10**
10. **20**	11. **99**	12. **49**
13. **81**	14. **1**	15. **95**
16. **79**	17. **301**	18. **77**
19. **28**	20. **412**	21. **9**
22. **119**	23. **153**	24. **1**
25. **184**	26. **7**	27. **49**
28. **130**	29. **401**	30. **2830**
31. **128**	32. **968**	33. **705**
34. **99**	35. **4161**	36. **963**
37. **1013**	38. **41,986**	39. **1589**
40. **321**	41. **887**	42. **6123**
43. **55,879**	44. **5503**	

45. **6492**	46. **4305**
47. **58,621**	48. **10,191**
49. **1,703,643**	50. **9449**

EXERCISE 9a (PAGE 51)

1. **24**	2. **24**	3. **80**	4. **80**
5. **72**	6. **72**	7. **35**	8. **35**
9. **55**	10. **55**	11. **24**	12. **24**
13. **18**	14. **18**	15. **14**	16. **14**
17. **40**	18. **40**	19. **132**	20. **132**
21. **0**	22. **0**	23. **0**	24. **0**
25. **0**	26. **0**	27. **0**	28. **0**
29. **0**	30. **0**	31. **5**	32. **4**
33. **3**	34. **8**	35. **2**	36. **9**
37. **7**	38. **3**	39. **8**	40. **6**

Word Problem

Thelma will need **24 plants** next season. (8 × 3 = 24)

EXERCISE 9b (PAGE 52)

1. **39**	2. **42**	3. **24**	4. **60**
5. **77**	6. **69**	7. **82**	8. **88**
9. **98**	10. **28**	11. **0**	12. **16**
13. **60**	14. **68**	15. **66**	16. **444**
17. **936**	18. **7299**	19. **999**	20. **333**
21. **909**	22. **228**	23. **0**	24. **999**
25. **448**			

Word Problem

Stacia paid **$36** for all the books. (12 × $3 = $36)

EXERCISE 9c (PAGE 53)

1. **112**	2. **114**	3. **52**
4. **84**	5. **96**	6. **64**
7. **75**	8. **78**	9. **85**
10. **438**	11. **812**	12. **1260**
13. **3105**	14. **5528**	15. **1494**
16. **832**	17. **1746**	18. **1713**
19. **3760**	20. **6660**	21. **20,620**
22. **21,224**	23. **9,648**	24. **43,610**
25. **20,660**		

Word Problem

Kay's total pay was **$15,400**. ($2200 × 7 = $15,400)

EXERCISE 9d (PAGE 55)

1. **504**　　2. **690**　　3. **420**
4. **880**　　5. **352**　　6. **484**
7. **190**　　8. **714**　　9. **610**
10. **492**　　11. **30,408**　　12. **16,472**
13. **7168**　　14. **18,648**　　15. **13,975**
16. **12,240**　　17. **19,264**　　18. **13,816**
19. **135,120**　　20. **523,600**　　21. **413,318**
22. **459,255**　　23. **17,952**　　24. **90,741**
25. **651,816**

Word Problem

Ellen spent a total of **$308**.
($11 \times $28 = 308)

EXERCISE 9e (PAGE 56)

1. **69,394**　　2. **103,683**　　3. **17,182**
4. **99,405**　　5. **58,646**　　6. **234,567**
7. **147,550**　　8. **133,416**　　9. **568,108**
10. **1,126,554**　　11. **747,585**
12. **3,973,673**　　13. **4,820,743**
14. **3,991,428**　　15. **255,673**
16. **296,800**　　17. **918,999**
18. **1,635,390**　　19. **2,101,908**
20. **1,877,037**

Word Problem

Marcia should be paid a total of **$108,108**
for the stoves. ($$468 \times 231 = $108,108$)

EXERCISE 10a (PAGE 58)

1. **420**　　2. **990**　　3. **4480**
4. **1760**　　5. **5280**　　6. **9360**
7. **4660**　　8. **5600**　　9. **17,940**
10. **23,380**　　11. **4040**　　12. **9120**
13. **37,920**　　14. **29,970**　　15. **21,330**
16. **63,630**　　17. **129,630**　　18. **135,780**
19. **63,050**　　20. **60,000**

Word Problem

The team ordered **8640 baseballs**
altogether. ($144 \times 60 = 8640$)

EXERCISE 10b (PAGE 59)

1. **63,336**　　2. **12,423**　　3. **92,862**
4. **167,862**　　5. **91,494**　　6. **127,072**
7. **85,023**　　8. **224,775**　　9. **666,398**
10. **700,536**　　11. **601,398**　　12. **319,656**
13. **1,897,405**　　14. **1,073,402**
15. **1,955,328**　　16. **2,553,684**

Word Problem

Mr. Whitman paid **$13,566** in all for the
pens. ($102 \times $133 = $13,566$)

EXERCISE 10c (PAGE 60)

1. **360**　　2. **720**　　3. **340**
4. **4040**　　5. **670**　　6. **8480**
7. **7650**　　8. **550**　　9. **300**
10. **34,500**　　11. **70,200**　　12. **41,900**
13. **789,000**　　14. **4,321,000**
15. **6,279,000**　　16. **316,000**　　17. **4230**
18. **4200**　　19. **914,000**　　20. **618,300**

Word Problem

The total weight of the boxes was **7500
pounds.** ($75 \times 100 = 7500$)

EXERCISE 11a (PAGE 63)

1. **7**　　2. **7**　　3. **2**　　4. **0**　　5. **6**
6. **1**　　7. **9**　　8. **3**　　9. **4**　　10. **3**
11. **6**　　12. **5**　　13. **5**　　14. **2**　　15. **7**
16. **3**　　17. **2**　　18. **3**　　19. **6**　　20. **0**
21. **8**　　22. **7**　　23. **9**　　24. **3**　　25. **9**
26. **8**　　27. **6**　　28. **1**　　29. **4**　　30. **9**
31. **7**　　32. **8**　　33. **8**　　34. **4**　　35. **7**
36. **2**　　37. **8**　　38. **5**　　39. **6**　　40. **9**
41. **5**　　42. **1**　　43. **0**　　44. **8**　　45. **4**
46. **7**　　47. **3**　　48. **0**　　49. **5**　　50. **2**
51. **8**　　52. **6**　　53. **5**　　54. **9**　　55. **8**
56. **6**　　57. **9**　　58. **5**　　59. **4**　　60. **7**
61. **0**　　62. **6**　　63. **2**　　64. **9**　　65. **5**
66. **3**　　67. **4**　　68. **5**　　69. **9**　　70. **7**
71. **1**　　72. **4**　　73. **1**　　74. **0**　　75. **9**
76. **6**　　77. **8**　　78. **3**　　79. **7**　　80. **2**

Word Problem

Each person contributed **$6** toward the
price of the cake. ($$24 \div 4 = 6)

EXERCISE 11b (PAGE 64)

1. **24**　　2. **21**　　3. **11**　　4. **32**
5. **6**　　6. **31**　　7. **22**　　8. **22**
9. **21**　　10. **32**　　11. **211**　　12. **122**
13. **234**　　14. **121**　　15. **212**　　16. **321**
17. **111**　　18. **100**　　19. **222**　　20. **212**
21. **1221**　　22. **2233**　　23. **1212**　　24. **4321**
25. **1111**

Word Problem

Each person got **$32**. ($$96 \div 3 = 32)

EXERCISE 11c (PAGE 65)

1. **51** 2. **83** 3. **62** 4. **92**
5. **91** 6. **62** 7. **31** 8. **82**
9. **93** 10. **81** 11. **31** 12. **41**
13. **91** 14. **72** 15. **83** 16. **411**
17. **412** 18. **711** 19. **523** 20. **723**
21. **311** 22. **822** 23. **812** 24. **911**
25. **311**

Word Problem

Charley put **82 books** in each box.
(328 ÷ 4 = 82)

EXERCISE 11d (PAGE 67)

1. **208** 2. **205** 3. **102** 4. **406**
5. **102** 6. **103** 7. **202** 8. **306**
9. **104** 10. **101** 11. **105** 12. **207**
13. **103** 14. **404** 15. **607** 16. **1010**
17. **2043** 18. **2201** 19. **1011** 20. **2072**
21. **1031** 22. **4103** 23. **1031** 24. **1021**
25. **1051**

Word Problem

The amount of each payment was **$2092**.
($6276 ÷ 3 = $2092)

EXERCISE 11e (PAGE 68)

1. **63** 2. **63** 3. **46** 4. **34**
5. **54** 6. **13** 7. **74** 8. **81**
9. **71** 10. **68** 11. **20** 12. **76**
13. **68** 14. **34** 15. **44** 16. **80**
17. **93** 18. **77** 19. **81**
20. **39** 21. **743** 22. **624**
23. **653** 24. **924** 25. **126**

Word Problem

There were **43 recruits** in each training
platoon. (258 ÷ 6 = 43)

EXERCISE 11f (PAGE 70)

1. **2 r1** 2. **2 r1** 3. **3 r1** 4. **1 r2**
5. **2 r1** 6. **7 r1** 7. **8 r3** 8. **4 r4**
9. **6 r2** 10. **6 r2** 11. **26 r1** 12. **4 r3**
13. **8 r3** 14. **4 r3** 15. **11 r1** 16. **16 r1**
17. **12 r3** 18. **12 r2** 19. **16 r1** 20. **15 r4**
21. **103 r2** 22. **277 r1** 23. **243 r1**
24. **125 r3** 25. **121 r2**

Word Problem

Each day the number of miles Ruben
drove was **328 r1**. (985 ÷ 3 = 328 r1)

EXERCISE 12a (PAGE 71)

1. **5** 2. **9 r4** 3. **8** 4. **3**
5. **12 r4** 6. **6** 7. **5** 8. **3**
9. **5 r8** 10. **2 r13** 11. **4** 12. **5**
13. **4** 14. **8** 15. **5** 16. **6 r9**
17. **9 r34** 18. **4 r6** 19. **6 r3** 20. **7**
21. **4 r5** 22. **4** 23. **7 r22** 24. **3 r23**
25. **7 r32**

Word Problem

There will be **21 parents** in each room.
(525 ÷ 25 = 21)

EXERCISE 12b (PAGE 73)

1. **211 r5** 2. **112** 3. **231 r1**
4. **321** 5. **435 r10** 6. **580**
7. **104** 8. **144 r25** 9. **293 r12**
10. **63** 11. **213** 12. **181 r14**
13. **113 r 21** 14. **48** 15. **212 r20**
16. **413** 17. **528** 18. **1365 r6**
19. **2222 r5** 20. **2200 r16** 21. **1160**
22. **334** 23. **1953 r12** 24. **526**
25. **1831**

Word Problem

In each theater, **125 people** saw the
movie. (4375 ÷ 35 = 125)

EXERCISE 12c (PAGE 74)

1. **3 r5** 2. **4 r4** 3. **2 r5**
4. **2 r10** 5. **3** 6. **4 r100**
7. **10 r140** 8. **31 r11** 9. **21 r181**
10. **21 r2** 11. **5** 12. **3 r19**
13. **21 r2** 14. **17 r188** 15. **17 r309**
16. **13 r90** 17. **32 r8** 18. **37 r208**
19. **2** 20. **20 r300** 21. **113 r80**
22. **54** 23. **111 r274** 24. **85 r120**

Word Problem

Each employee's bonus was **$196**.
($23,520 ÷ 120 = $196)

MIXED PRACTICE 2 (page 76)

1. **168** 2. **8** 3. **48**
4. **1** 5. **0** 6. **8**
7. **7** 8. **7** 9. **42**
10. **20** 11. **122** 12. **1002**
13. **4038** 14. **1428** 15. **202**
16. **14,046** 17. **1111** 18. **212**

19. **21,392**   20. **13,000**   21. **432**
22. **314**   23. **311**   24. **621**
25. **1240**   26. **16,096**   27. **6048**
28. **801**   29. **403**   30. **924**
31. **133,108**   32. **333,321**   33. **4 r3**
34. **9 r1**   35. **4 r2**   36. **11 r7**
37. **183,425**   38. **148,862**   39. **20 r5**
40. **202**   41. **534,950**   42. **308,308**
43. **1,258,136**   44. **102**   45. **104 r16**
46. **1,801,800**   47. **4,079,769**   48. **359 r9**
49. **0**   50. **2004 r13**   51. **2**
52. **6 r322**   53. **443,232**
54. **1,488,000**   55. **18,952,880**
56. **2,268,721**   57. **101 r80**
58. **68 r206**   59. **1,744,002**
60. **158 r55**

WHOLE-NUMBER SKILLS REVIEW (page 77)

Part A

1. **216**   2. **38**   3. **94**
4. **369**   5. **3808**   6. **1436**
7. **9490**   8. **3**   9. **1832**
10. **1591**   11. **23**   12. **420**
13. **25**   14. **32,778**   15. **5333**
16. **54 r7**   17. **16,842**   18. **392**
19. **750**   20. **466**   21. **287**
22. **1837**   23. **14,460**   24. **30**
25. **491**   26. **250,848**   27. **5**
28. **565**   29. **23**   30. **26,700**

Part B

1. Denise should pack **16 bags** in each box.
   (384 ÷ 24 = 16)
2. Esteban earned **$647** in all.
   ($478 + $169 = $647)
3. Carmella inspected **408 more blouses** on Monday than on Tuesday.
   (1746 − 2154 = 408)
4. Altogether, Chan paid **$432** for the teacups. ($12 × 36 = $432)
5. There will be **3 roses** in each centerpiece.
   (183 ÷ 61 = 3)
6. The lumber yard spends **$189,600** each year for their salaries.
   ($15,800 × 12 = $189,600)
7. Henry drove **855 mi** in all.
   (227 + 341 + 287 = 855)
8. The amount of Henry's pay was **$184** after taxes. ($240 − $56 = $184)

EXERCISE 13a (PAGE 79)

1. **108 in.**   2. **4000 mm**   3. **5280 yd**
4. **70 m**   5. **3888 in.**   6. **10 m**
7. **2000 m**   8. **129 ft**   9. **30 hm**
10. **144 in.**

Word Problem

    Yes, the board Peter has is long enough to serve for the shelf. (4 ft × 12 = 48 in.)

EXERCISE 13b (PAGE 80)

1. **1 yd**   2. **4 m**   3. **15 ft**
4. **43 yd**   5. **3 km**   6. **10 yd**
7. **4 km**   8. **2 mi**   9. **8 hm**
10. **6 dam**

Word Problem

    The snake is **9 feet** long.
(108 in. ÷ 12 = 9 ft)

EXERCISE 13c (PAGE 81)

1. **4000 lb**   2. **2000 mg**   3. **6 lb**
4. **192 oz**   5. **2 kg**   6. **5 T**
7. **2 lb**   8. **10,000 lb**   9. **30 dg**
10. **4 lb**   11. **400 cg**   12. **8 T**

Word Problem

    No, Jason would not have enough mushrooms because he needs 16 ounces of mushrooms. (1 lb × 16 = 16 oz)

EXERCISE 13d (PAGE 82)

1. **80 fl oz**   2. **4000 ml**   3. **2 gal**
4. **32 qt**   5. **8 gal**   6. **3000 hl**
7. **22 pt**   8. **200 L**   9. **4 qt**
10. **2 gal**   11. **16 C**   12. **5 L**

Word Problem

    Kerry should buy **a pint** container of milk every day. (16 oz ÷ 16 = 1 pt)

EXERCISE 13e (PAGE 83)

1. **180 sec**   2. **6 wk**   3. **280 da**
4. **4 min**   5. **240 min**   6. **15 yr**
7. **2 yr**   8. **36 mo**   9. **3600 sec**
10. **4 yr**

Word Problem

    Mildred is likely to be sick for **3 days.**
(72 hr ÷ 24 = 3 da)

Part A

1. **10,560 ft**		2. **6 yd**	
3. **700 daL**		4. **16 hr**	
5. **3 mi**		6. **5 dam**	
7. **5 yr**		8. **200 cg**	
9. **144 in.**		10. **11 mi**	
11. **40 hg**		12. **10 yr**	
13. **6 ft**		14. **216 in.**	
15. **4 lb**		16. **4000 lb**	
17. **364 da**		18. **3 da**	
19. **16 lb**		20. **8 T**	
21. **32 pt**		22. **2 qt**	
23. **96 C**		24. **80 dL**	
25. **2000 mg**			

Part B

1. Andrea must buy **16 qt** of oil.
   (4 gal × 4 = 16 qt)

2. Georgia needs **8** quart containers.
   (16 pt ÷ 2 = 8 qt)

3. Susan bought **468 in.** of ribbon.
   (13 yd × 36 = 468 in.)

4. Fred bought **24 yd** of tape.
   (864 in ÷ 36 = 24 yd)

5. Alberto was away for **49 da.**
   (7 wk × 7 = 49 da)

EXERCISE 14a (PAGE 85)

1. **11 yd 2 ft**		2. **10 hr 50 min**	
3. **6 T 1600 lb**		4. **10 gal 3 qt**	
5. **46 min 18 sec**		6. **21 lb 7 oz**	
7. **33 mi 442 yd**		8. **20 mo 3 wk**	
9. **21 T 409 lb**		10. **61 qt 1 pt**	
11. **101 gal 3 qt**		12. **15 qt 31 fl oz**	

Word Problem

Harry worked out **59 min 49 sec** that day.

$$
\begin{array}{r}
35 \text{ min } 38 \text{ sec} \\
+\ 24 \text{ min } 11 \text{ sec} \\
\hline
59 \text{ min } 49 \text{ sec}
\end{array}
$$

EXERCISE 14b (PAGE 87)

Each problem is solved using the methods shown in both Examples 2 and 3.

1.

Example 2 Method

(Step 1)
$$
\begin{array}{r}
15 \text{ min } 45 \text{ sec} \\
+\ 20 \text{ min } 20 \text{ sec} \\
\hline
35 \text{ min } 65 \text{ sec}
\end{array}
$$

(Step 2)
$$
\begin{array}{r}
1 \text{ min } 5 \text{ sec} \\
60 \overline{)\ 65 \text{ sec}} \\
\underline{60} \\
5
\end{array}
$$

(Step 3)
$$
\begin{array}{r}
35 \text{ min} \\
+\ 1 \text{ min } 5 \text{ sec} \\
\hline
\mathbf{36 \text{ min } 5 \text{ sec}}
\end{array}
$$

Example 3 Method

(Step 1)
$$
\begin{array}{r}
15 \text{ min} \\
\times\ 60 \\
\hline
900 \text{ sec} \\
+\ 45 \text{ sec} \\
\hline
945 \text{ sec}
\end{array}
$$

(Step 2)
$$
\begin{array}{r}
20 \text{ min} \\
\times\ 60 \\
\hline
1200 \text{ sec} \\
+\ 20 \text{ sec} \\
\hline
1220 \text{ sec}
\end{array}
$$

(Step 3)
$$
\begin{array}{r}
945 \text{ sec} \\
+\ 1220 \text{ sec} \\
\hline
2165 \text{ sec}
\end{array}
$$

(Step 4)
$$
\begin{array}{r}
36 \text{ min } 5 \text{ sec} \\
60 \overline{)\ 2165 \text{ sec}} \\
\underline{180} \\
365 \\
\underline{360} \\
5
\end{array}
$$

2.

Example 2 Method

(Step 1)
$$
\begin{array}{r}
16 \text{ lb } 10 \text{ oz} \\
+\ 20 \text{ lb } 12 \text{ oz} \\
\hline
36 \text{ lb } 22 \text{ oz}
\end{array}
$$

(Step 2)
$$
\begin{array}{r}
1 \text{ lb } 6 \text{ oz} \\
16 \overline{)\ 22 \text{ oz}} \\
\underline{16} \\
6
\end{array}
$$

(Step 3)
$$
\begin{array}{r}
36 \text{ lb} \\
+\ 1 \text{ lb } 6 \text{ oz} \\
\hline
\mathbf{37 \text{ lb } 6 \text{ oz}}
\end{array}
$$

Example 3 Method

(Step 1)
$$
\begin{array}{r}
16 \text{ lb} \\
\times\ 16 \\
\hline
256 \text{ oz} \\
+\ 10 \text{ oz} \\
\hline
266 \text{ oz}
\end{array}
$$

(Step 2)
$$
\begin{array}{r}
20 \text{ lb} \\
\times\ 16 \\
\hline
320 \text{ oz} \\
+\ 12 \text{ oz} \\
\hline
332 \text{ oz}
\end{array}
$$

(Step 3)
$$
\begin{array}{r}
266 \text{ oz} \\
+\ 332 \text{ oz} \\
\hline
598 \text{ oz}
\end{array}
$$

(Step 4)
$$
\begin{array}{r}
37 \text{ lb } 6 \text{ oz} \\
16 \overline{)\ 598 \text{ oz}} \\
\underline{48} \\
118 \\
\underline{112} \\
6
\end{array}
$$

3.

Example 2 Method

(Step 1)
$$
\begin{array}{r}
5 \text{ qt } 1 \text{ pt} \\
+\ 9 \text{ qt } 1 \text{ pt} \\
\hline
14 \text{ qt } 2 \text{ pt}
\end{array}
$$

(Step 2)
$$
\begin{array}{r}
1 \text{ qt} \\
2 \overline{)\ 2 \text{ pt}} \\
\underline{2} \\
0
\end{array}
$$

(Step 3)
$$
\begin{array}{r}
14 \text{ qt} \\
+\ 1 \text{ qt} \\
\hline
\mathbf{15 \text{ qt}}
\end{array}
$$

Example 3 Method

(Step 1)
```
    5 qt
×    2
  10 pt
+  1 pt
  11 pt
```
(Step 2)
```
    9 qt
×    2
  18 pt
+  1 pt
  19 pt
```

(Step 3)
```
   11 pt
+  19 pt
   30 pt
```
(Step 4)
```
        15 qt
    2) 30 pt
        2
        10
        10
         0
```

4. Example 2 Method

(Step 1)
```
   3 ft   8 in.
+  6 ft   6 in.
   9 ft  14 in.
```
(Step 2)
```
        1 ft 2 in.
    12) 14 in.
        12
         2
```

(Step 3)
```
   9 ft
+  1 ft 2 in.
  10 ft 2 in.
```

Example 3 Method

(Step 1)
```
      12
×   3 ft
   36 in.
+   8 in.
   44 in.
```
(Step 2)
```
      12
×   6 ft
   72 in.
+   6 in.
   78 in.
```

(Step 3)
```
   44 in.
+  78 in.
  122 in.
```
(Step 4)
```
          10 ft 2 in.
    12) 122 in.
        12
         2
         0
         2
```

5. Example 2 Method

(Step 1)
```
    65 lb 14 oz
+   40 lb 12 oz
   105 lb 26 oz
```
(Step 2)
```
         1 lb 10 oz
    16) 26 oz
        16
        10
```

(Step 3)
```
   105 lb
+    1 lb 10 oz
   106 lb 10 oz
```

Example 3 Method

(Step 1)
```
     65 lb
×      16
   1040 oz
+    14 oz
   1054 oz
```
(Step 2)
```
     40 lb
×      16
    640 oz
+    12 oz
    652 oz
```

(Step 3)
```
   1054 oz
+   652 oz
   1706 oz
```
(Step 4)
```
          106 lb 10 oz
    16) 1706 oz
         16
        106
         96
         10
```

6. Example 2 Method

(Step 1)
```
    3 C  6 fl oz
+   8 C  4 fl oz
   11 C 10 fl oz
```
(Step 2)
```
        1 C 2 fl oz
    8) 10 fl oz
        8
        2
```

(Step 3)
```
   11 C
+   1 C 2 fl oz
   12 C 2 fl oz
```

Example 3 Method

(Step 1)
```
       3 C
×      8
   24 fl oz
+   6 fl oz
   30 fl oz
```
(Step 2)
```
       8 C
×      8
   64 fl oz
+   4 fl oz
   68 fl oz
```

(Step 3)
```
   30 fl oz
+  68 fl oz
   98 fl oz
```
(Step 4)
```
          12 C 2 fl oz
    8) 98 fl oz
        8
        18
        16
         2
```

7. Example 2 Method

(Step 1)
```
   12 wk 6 da
+   9 wk 3 da
   21 wk 9 da
```
(Step 2)
```
        1 wk 2 da
    7) 9 da
        7
        2
```

(Step 3)
```
   21 wk
+   1 wk 2 da
   22 wk 2 da
```

Example 3 Method

(Step 1)
```
     12 wk
×      7
   84 da
+   6 da
   90 da
```
(Step 2)
```
      9 wk
×      7
   63 da
+   3 da
   66 da
```

(Step 3)
```
   90 da
+  66 da
  156 da
```
(Step 4)
```
          22 wk 2 da
    7) 156 da
        14
        16
        14
         2
```

8.

Example 2 Method

(Step 1)
```
   321 T 1840 lb
+   99 T  605 lb
   420 T 2445 lb
```

(Step 2)
```
           1 T 445 lb
2000 ) 2445 lb
       2000
        445
```

(Step 3)
```
   420 T
+    1 T 445 lb
   421 T 445 lb
```

Example 3 Method

(Step 1)
```
      321 T
×     2000
   642,000 lb
+    1 840 lb
   643,840 lb
```

(Step 2)
```
       99 T
×    2000
   198,000 lb
+      605 lb
   198,605 lb
```

(Step 3)
```
   643,840 lb
+ 198,605 lb
   842,445 lb
```

(Step 4)
```
            421 T 445 lb
2000 ) 842,445 lb
       800 0
        42 44
        40 00
         2 445
         2 000
           445
```

9.

Example 2 Method

(Step 1)
```
   45 min 32 sec
+   5 min 50 sec
   50 min 82 sec
```

(Step 2)
```
        1 min 22 sec
60 ) 82 sec
     60
     22
```

(Step 3)
```
   50 min
+   1 min 22 sec
   51 min 22 sec
```

Example 3 Method

(Step 1)
```
     45 min
×     60
   2700 sec
+    32 sec
   2732 sec
```

(Step 2)
```
      5 min
×     60
    300 sec
+    50 sec
    350 sec
```

(Step 3)
```
   2732 sec
+   350 sec
   3082 sec
```

(Step 4)
```
         51 min 22 sec
60 ) 3082 sec
     300
      82
      60
      22
```

10.

Example 2 Method

(Step 1)
```
    6 yr  9 mo
+   5 yr  4 mo
   11 yr 13 mo
```

(Step 2)
```
         1 yr 1 mo
12 ) 13 mo
     12
      1
```

(Step 3)
```
   11 yr
+   1 yr 1 mo
   12 yr 1 mo
```

Example 3 Method

(Step 1)
```
      6 yr
×    12
   72 mo
+   9 mo
   81 mo
```

(Step 2)
```
      5 yr
×    12
   60 mo
+   4 mo
   64 mo
```

(Step 3)
```
    81 mo
+  64 mo
   145 mo
```

(Step 4)
```
         12 yr 1 mo
12 ) 145 mo
     12
      25
      24
       1
```

11.

Example 2 Method

(Step 1)
```
    2 pt 1 C
+   8 pt 1 C
   10 pt 2 C
```

(Step 2)
```
      1 pt
2 ) 2 C
    2
    0
```

(Step 3)
```
   10 pt
+   1 pt
   11 pt
```

Example 3 Method

(Step 1)
```
    2 pt
×  2
   4 C
+  1 C
   5 C
```

(Step 2)
```
    8 pt
×   2
   16  C
+   1  C
   17  C
```

(Step 3)
```
    5 C
+  17 C
   22 C
```

(Step 4)
```
       11 pt
2 ) 22 C
    2
    2
    2
    0
```

12.

Example 2 Method

(Step 1)   12 min 42 sec
           + 9 min 51 sec
           21 min 93 sec

(Step 2)   1 min 33 sec
          60) 93 sec
               60
               33

(Step 3)   21 min
           + 1 min 33 sec
           22 min 33 sec

Example 3 Method

(Step 1)   12 min   (Step 2)   9 min
          × 60   × 60
          720 sec   540 sec
         + 42 sec   + 51 sec
          762 sec   591 sec

(Step 3)   762 sec   (Step 4)   **22 min 33 sec**
            591 sec   60) 1353 sec
           1353 sec   120
                                    153
                                    120
                                     33

Word Problem

Charles worked 56 hr per week.

    40 hr 30 min
  + 15 hr 30 min
    55 hr 60 min, or 56 hr

EXERCISE 14c (PAGE 88)

All problems are solved using the method shown in Example 4. Problems that require borrowing are also solved using the method shown in Example 5.

1.   Example 4 Method

    10 gal 3 qt
  − 7 gal 1 qt
    3 gal 2 qt

2.   Example 4 Method

       14   7
      1̶5̶ wk 0̶ da
    − 6 wk 3 da
     8 wk 4 da

Example 5 Method

(Step 1)   15 wk   (Step 2)   6 wk
          ×   7   × 7
           105 da   42 da
                                  + 3 da
                                    45 da

(Step 3)   105 da   (Step 4)   **8 wk 4 da**
          − 45 da   7) 60 da
            60 da   56
                                     4

3.   Example 4 Method

    12 T 500 lb
  − 8 T 300 lb
    4 T 200 lb

4.   Example 4 Method

     8 yd 2 ft
    −6 yd 1 ft
    2 yd 1 ft

5.   Example 4 Method

      3   20
     4̶ lb   4̶ oz
    −2 lb   8 oz
    1 lb 12 oz

Example 5 Method

(Step 1)   16   (Step 2)   16
          × 4 lb   ×   2 lb
           64 oz   32 oz
         + 4 oz   + 8 oz
           68 oz   40 oz

(Step 3)   68 oz   (Step 4)   **1 lb 12 oz**
          − 40 oz   16) 28 oz
            28 oz   16
                                   12

6.   Example 4 Method

      29   70
     3̶0̶ min 1̶0̶ sec
    − 18 min 25 sec
    11 min 45 sec

Example 5 Method

(Step 1)   30 min   (Step 2)   18 min
          × 60   × 60
           1800 sec   1080 sec
         +   10 sec   +   25 sec
           1810 sec   1105 sec

(Step 3)   1810 sec   (Step 4)   **11 min 45 sec**
           −1105 sec   60) 705 sec
            705 sec   60
                                    105
                                     60
                                     45

7.

<u>Example 4 Method</u>

```
  36 ft 9 in.
− 20 ft 9 in.
  16 ft
```

8.

<u>Example 4 Method</u>

```
   9    34
  1̶0̶ qt  2̶ fl oz
−  6 qt  4 fl oz
   3 qt 30 fl oz
```

<u>Example 5 Method</u>

(Step 1)
```
        32
   ×  10 qt
     320 fl oz
   +   2 fl oz
     322 fl oz
```

(Step 2)
```
        32
   ×    6
     192 fl oz
   +   4 fl oz
     196 fl oz
```

(Step 3)
```
     322 fl oz
   − 196 fl oz
     126 fl oz
```

(Step 4)
```
        3 qt 30 fl oz
   32) 126 fl oz
        96
        30
```

9.

<u>Example 4 Method</u>

```
   15    16
  1̶6̶ lb  0̶ oz
−  4 lb  6 oz
   11 lb 10 oz
```

<u>Example 5 Method</u>

(Step 1)
```
        16
   ×  16 lb
     256 oz
```

(Step 2)
```
        16
   +   4 lb
      64 oz
   +   6 oz
      70 oz
```

(Step 3)
```
     256 oz
   −  70 oz
     186 oz
```

(Step 4)
```
        11 lb 10 oz
   16) 186 oz
        16
        26
        16
        10
```

10.

<u>Example 4 Method</u>

```
   2     1760
  3̶ mi   0̶ yd
− 0 mi 1500 yd
  2 mi  260 yd
```

<u>Example 5 Method</u>

(Step 1)
```
     1760
   ×   3 mi
     5280 yd
```

(Step 2)   1500 yd

(Step 3)
```
  5280 yd
− 1500 yd
  3780 yd
```

(Step 4)
```
        2 mi 260 yd
  1760) 3780 yd
         3520
          260
```

11.

<u>Example 4 Method</u>

```
   45    22
  4̶6̶ ft 1̶0̶ in.
−  8 ft 11 in.
   37 ft 11 in.
```

<u>Example 5 Method</u>

(Step 1)
```
        46 ft
   ×    12
       552 in.
   +   10 in.
       562 in.
```

(Step 2)
```
        12
   ×   8 ft
      96 in.
   +  11 in.
     107 in.
```

(Step 3)
```
     562 in.
   − 107 in.
     455 in.
```

(Step 4)
```
        37 ft 11 in.
   12) 455 in.
        36
        95
        84
        11
```

12.

<u>Example 4 Method</u>

```
   54     74
  5̶5̶ min 1̶4̶ sec
− 16 min 20 sec
  38 min 54 sec
```

<u>Example 5 Method</u>

(Step 1)
```
        55 min
   ×    60
      3300 sec
   +    14 sec
      3314 sec
```

(Step 2)
```
        16 min
   ×    60
       960 sec
   +    20 sec
       980 sec
```

(Step 3)
```
      3314 sec
   −   980 sec
      2334 sec
```

(Step 4)
```
         38 min 54 sec
   60) 2334 sec
        180
        534
        480
         54
```

Word Problem

The weight of the onions left in the sack was **7 lb 12 oz.**

<u>Example 4 Method</u>

```
   9    24
  1̶0̶ lb  8̶ oz
−  2 lb 12 oz
   7 lb 12 oz
```

Example 5 Method

(Step 1)
```
       16
   ×  10 lb
     160 oz
   +   8 oz
     168 oz
```

(Step 2)
```
       16
   ×   2 lb
      32 oz
   + 12 oz
      44 oz
```

(Step 3)
```
     168 oz
   −  44 oz
     124 oz
```

(Step 4)
```
         7 lb 12 oz
   16) 124 oz
       112
        12
```

EXERCISE 14d (PAGE 89)

1. **24 T 1600 lb**
2. **12 da 18 hr**
3. **30 ft 10 in.**
4. **36 qt 24 fl oz**
5. **77 lb 14 oz**
6. **9 mi 660 yd**
7. **32 yd 2 ft**
8. **420 mi 3 160 ft**
9. **185 min 55 sec**
10. **168 lb 12 oz**
11. **24 qt 12 fl oz**
12. **54 T 882 lb**

Word Problem

Alicia needed **84 ft 6 in.** of nylon line.

```
    14 ft 1 in.
   ×       6
    84 ft 6 in.
```

EXERCISE 14e (PAGE 91)

Each problem is solved using the methods shown in both Examples 7 and 8.

1.
Example 7 Method

(Step 1)
```
    1 mi   440 yd
  ×          4
   4 mi 1760 yd
```

(Step 2)
```
              1 mi
  1760) 1760 yd
         1760
            0
```

(Step 3)
```
     4 mi
   + 1 mi
     5 mi
```

Example 8 Method

(Step 1)
```
       1760
   ×      1 mi
       1760 yd
   +    440 yd
       2200 yd
```

(Step 2)
```
      2200 yd
   ×       4
      8800 yd
```

(Step 3)
```
              5 mi
   1760) 8800 yd
          8800
             0
```

2.
Example 7 Method

(Step 1)
```
    4 lb 12 oz
  ×        8
   32 lb 96 oz
```

(Step 2)
```
            6 lb
  16) 96 oz
        96
         0
```

(Step 3)
```
    32 lb
   + 6 lb
    38 lb
```

Example 8 Method

(Step 1)
```
       16
       4 lb
      64 oz
   + 12 oz
      76 oz
```

(Step 2)
```
      76 oz
   ×    8
     608 oz
```

(Step 3)
```
          38 lb
   16) 608 oz
        48
       128
       128
         0
```

3.
Example 7 Method

(Step 1)
```
    14 pt   8 oz
  ×         6
   84 pt 48 oz
```

(Step 2)
```
            3 pt
  16) 48 oz
        48
         0
```

(Step 3)
```
    84 pt
   + 3 pt
    87 pt
```

Example 8 Method

(Step 1)
```
       14 pt
   ×   16
       84
       14
      224 oz
   +   8 oz
      232 oz
```

(Step 2)
```
      232 oz
   ×     6
     1392 oz
```

(Step 3)
```
            87 pt
   16) 1392 oz
        128
        112
        112
          0
```

4.
Example 7 Method

(Step 1)
```
    72 min 23 sec
  ×          4
   288 min 92 sec
```

(Step 2)
```
          1 min 32 sec
  60 ) 92 sec
        60
        32
```

(Step 3)
```
    288 min
   +   1 min 32 sec
    289 min 32 sec
```

Example 8 Method

(Step 1)
```
      72 min
  ×     60
    4320  sec
  +    23  sec
    4343  sec
```

(Step 2)
```
    4343 sec
  ×      4
   17372 sec
```

(Step 3)
```
        289 min 32 sec
    60) 17372 sec
        120
        ────
        537
        480
        ────
        572
        540
        ────
         32
```

5.
Example 7 Method

(Step 1)
```
     12 qt   1 pt
   ×       23
    276 qt 23 pt
```

(Step 2)
```
           11 qt 1 pt
      2) 23 pt
         2
         ──
         3
         2
         ──
         1
```

(Step 3)
```
    276 qt
  +  11 qt 1 pt
    287 qt 1 pt
```

Example 8 Method

(Step 1)
```
     12 qt
   ×    2
     24 pt
   +  1 pt
     25 pt
```

(Step 2)
```
       25 pt
   ×    23
       75
       50
      ────
      575 pt
```

(Step 3)
```
        287 qt 1 pt
    2) 575 pt
       4
       ──
       17
       16
       ──
        15
        14
        ──
         1
```

6.
Example 7 Method

(Step 1)
```
      4 yr   8 mo
    ×        9
     36 yr 72 mo
```

(Step 2)
```
            6 yr
     12) 72 mo
         72
         ──
          0
```

(Step 3)
```
     36 yr
   +  6 yr
     42 yr
```

Example 8 Method

(Step 1)
```
        12
    ×  4 yr
       48 mo
    +   8 mo
       56 mo
```

(Step 2)
```
       56 mo
   ×      9
      504 mo
```

(Step 3)
```
          42 yr
     12) 504 mo
         48
         ──
         24
         24
         ──
          0
```

7.
Example 7 Method

(Step 1)
```
     32 wk   5 da
   ×         4
    128 wk 20 da
```

(Step 2)
```
            2 wk 6 da
      7) 20 da
         14
         ──
          6
```

(Step 3)
```
     128 wk
   +   2 wk 6  da
     130 wk 6 da
```

Example 8 Method

(Step 1)
```
      32 wk
    ×    7
     224  da
   +   5  da
     229  da
```

(Step 2)
```
       229 da
   ×      4
       916 da
```

(Step 3)
```
        130 wk 6 da
     7) 916
        7
        ──
        21
        21
        ──
         6
```

8.
Example 7 Method

(Step 1)
```
     16 C   4 fl oz
   ×        21
    336 C 84 fl oz
```

(Step 2)
```
            10 C 4 fl oz
      8) 84 fl oz
         8
         ──
         4
         0
         ──
         4
```

(Step 3)
```
     336 C
   +  10 C 4 fl oz
     346 C 4 fl oz
```

Example 8 Method

(Step 1)
```
        16 C
    ×    8
      128 fl oz
   +    4 fl oz
      132 fl oz
```

(Step 2)
```
       132 fl oz
   ×      21
       132
       264
      ─────
      2772 fl oz
```

(Step 3)
```
        346 C 4 fl oz
     8) 2772 fl oz
        24
        ──
        37
        32
        ──
        52
        48
        ──
         4
```

9. ───────── Example 7 Method ─────────

(Step 1)
```
   7 yd   28 in.
 ×         4
 ────────────────
 28 yd 112 in.
```
(Step 2)
```
        3 yd 4 in.
    36) 112 in.
        108
        ───
          4
```

(Step 3)
```
   28 yd
 +  3 yd 4 in.
 ─────────────
   31 yd 4 in.
```

───────── Example 8 Method ─────────

(Step 1)
```
      36
 ×     7 yd
 ──────────
    252 in.
 +   28 in.
 ──────────
    280 in.
```
(Step 2)
```
    280 in.
 ×      4
 ──────────
   1120 in.
```

(Step 3)
```
         31 yd 4 in.
    36) 1120
        108
        ───
         40
         36
         ──
          4
```

10. ───────── Example 7 Method ─────────

(Step 1)
```
   3 hr  24 min
 ×        3
 ───────────────
   9 hr  72 min
```
(Step 2)
```
        1 hr 12 min
    60) 72 min
        60
        ──
        12
```

(Step 3)
```
     9 hr
 +   1 hr 12 min
 ───────────────
  10 hr 12 min
```

───────── Example 8 Method ─────────

(Step 1)
```
      60
       3 hr
 ──────────
     180 min
 +    24 min
 ──────────
     204 min
```
(Step 2)
```
     204 min
 ×      3
 ──────────
     612 min
```

(Step 3)
```
         10 hr 12 min
    60) 612 min
        60
        ──
        12
         0
        ──
        12
```

11. ───────── Example 7 Method ─────────

(Step 1)
```
   24 mi      500 yd
 ×             32
 ──────────────────────
 768 mi 16,000 yd
```

(Step 2)
```
           9 mi 160 yd
  1760) 16,000 yd
        15,840
        ──────
           160
```

(Step 3)
```
    768 mi
 +    9 mi 160 yd
 ────────────────
  777 mi 160 yd
```

───────── Example 8 Method ─────────

(Step 1)
```
      1760
 ×      24 mi
 ──────────
      7040
      3520
      ──────
     42240 yd
 +     500 yd
 ──────────
     42740 yd
```
(Step 2)
```
     42 740 yd
 ×        32
 ──────────
     85 480
   1 282 20
 ──────────
  1,367,680 yd
```

(Step 3)
```
            777 mi 160 yd
  1760) 1,367,680
        1 232 0
        ───────
          135 68
          123 20
          ──────
           12 480
           12 320
           ──────
              160
```

12. ───────── Example 7 Method ─────────

(Step 1)
```
   32 gal   3 qt
 ×           7
 ────────────────
 224 gal 21 qt
```
(Step 2)
```
         5 gal 1 qt
    4) 21 qt
       20
       ──
        1
```

(Step 3)
```
   224 gal
 +   5 gal 1 qt
 ──────────────
   229 gal 1 qt
```

───────── Example 8 Method ─────────

(Step 1)
```
      32 gal
 ×     4
 ──────────
    128 qt
 +    3 qt
 ──────────
    131 qt
```
(Step 2)
```
     131 qt
 ×     7
 ──────────
    917 qt
```

(Step 3)
```
         229 gal 1 qt
    4) 917 qt
        8
        ──
        11
         8
        ──
        37
        36
        ──
         1
```

Word Problem

During those 7 weeks, George worked **311 hr 30 min** in all.

Example 7 Method

(Step 1)
```
   44 hr  30 min
 ×        7
  308 hr 210 min
```

(Step 2)
```
         3 hr 30 min
  60 ) 210 min
       180
        30
```

(Step 3)
```
  308 hr
 +  3 hr 30 min
  311 hr 30 min
```

Example 8 Method

(Step 1)
```
    44 hr
 ×  60
  2640 min
 +  30 min
  2670 min
```

(Step 2)
```
  2 670 min
 ×      7
  18,690 min
```

(Step 3)
```
         311 hr 30 min
  60 ) 18,690
       18 0
          69
          60
          90
          60
          30
```

EXERCISE 14f (PAGE 92)

1. **3 lb 2 oz**   2. **7 wk 2 da**   3. **6 yd 2 in.**
4. **4 gal 3 fl oz**   5. **7 lb 3 oz**   6. **6 ft 1 in.**
7. **21 mo 1 wk**   8. **15 hr 5 min**   9. **8 ft 3 in.**
10. **3 T 6 lb**   11. **4 da 2 min**   12. **38 gal 1 qt**

Word Problem

After the cuts, each board was **6 ft 3 in.** long.

```
      6 ft 3 in.
  3 ) 18 ft 9 in.
```

EXERCISE 14g (PAGE 93)

1. (Step 1)
```
    12
 ×  9 ft
    81 in.
 +   4 in.
    85 in.
```

(Step 2)
```
         17 in.
  5 ) 85 in.
       5
       35
       35
        0
```

(Step 3)
```
        1 ft 5 in.
  12 ) 17 in.
       12
        5
```

2. (Step 1)
```
    17 wk
 ×     7
   119 da
 +   1 da
   120 da
```

(Step 2)
```
         60 da
  2 ) 120 da
       12
       00
        0
        0
```

(Step 3)
```
        8 wk 4 da
  7 ) 60 da
      56
       4
```

3. (Step 1)
```
     16
     11 pt
     16
     16
    176 fl oz
 +   4 fl oz
    180 fl oz
```

(Step 2)
```
          18 fl oz
  10 ) 180 fl oz
        10
        80
        80
         0
```

(Step 3)
```
         1 pt 2 fl oz
  16 ) 18 fl oz
       16
        2
```

4. (Step 1)
```
     16 T
 ×  2,000
   32,000 lb
 +    500 lb
   32,500 lb
```

(Step 2)
```
         6 500 lb
  5 ) 32,500 lb
       30
        2 5
        2 5
        000
          0
          0
```

(Step 3)
```
            3 T 500 lb
  2000 ) 6500 lb
         6000
          500
```

5. (Step 1)
```
    22 gal
 ×   4
    88 qt
 +   2 qt
    90 qt
```

(Step 2)
```
         15 qt
  6 ) 90 qt
       6
       30
       30
        0
```

(Step 3)
```
        3 gal 3 qt
  4 ) 15 qt
      12
       3
```

6. (Step 1)
```
    14 lb
 ×  16
    84
    14
   224 oz
 +   6 oz
   230 oz
```

(Step 2)
```
         46 oz
  5 ) 230 oz
       20
       30
       30
        0
```

(Step 3)   **2 lb 14 oz**
16) 46 oz
    32
    14

7. (Step 1)   1760
         ×   7 mi
            12,320 yd
         +   181 yd
            12,501 yd

   (Step 2)   4 167 yd
            3) 12,501 yd
               12
                5
                3
               20
               18
               21
               21
                0

   (Step 3)   **2 mi 647 yd**
            1760) 4167 yd
                  3520
                   647

8. (Step 1)   23 mo
         ×   30
             690 da
         +   12 da
             702 da

   (Step 2)   78 da
            9) 702 da
               63
               72
               72
                0

   (Step 3)   **2 mo 12 da**
            30) 72 da
                60
                12

9. (Step 1)   10 pt
         ×   2
             20 C
         +   1 C
             21 C

   (Step 2)   7 C
            3) 21 C
                21
                 0

   (Step 3)   **3 pt 1 C**
            2) 7 C
               6
               1

10. (Step 1)   1760
                 9 mi
             15,840 yd
         +   281 yd
             16,121 yd

    (Step 2)   2 303 yd
            7) 16,121 yd
               14
                2 1
                2 1
                  2
                  0
                 21

    (Step 3)   **1 mi 543 yd**
            1760) 2303 yd
                  1760
                   543

11. (Step 1)   15 C
         ×   8
             120 fl oz
         +   4 fl oz
             124 fl oz

    (Step 2)   31 fl oz
            4) 124 fl oz
               12
                4
                4
                0

    (Step 3)   **3 C 7 fl oz**
            8) 31 fl oz
               24
                7

12. (Step 1)   12
         ×   10 yr
             120 mo
         +   8 mo
             128 mo

    (Step 2)   16 mo
            8) 128 mo
                8
               48
               48
                0

    (Step 3)   **1 yr 4 mo**
            12) 16 mo
                12
                 4

Word Problem

Sam ran **7 mi 644 yd** each day.

    (Step 1)   1760
         ×   22 mi
             3520
             3520
             38720 yd
         +   232 yd
             38952 yd

    (Step 2)   12,984 yd
            3) 38,952 yd
               3
               8
               6
               29
               27
               25
               24
               12
               12
                0

    (Step 3)   **7 mi 664 yd**
            1760) 12,984 yd
                   12 320
                      664

EXERCISE 15a (PAGE 95)

Part A

1. **20 ft**   2. **86 yd**
3. **104 in.**   4. **110 in.**
5. **120 ft**   6. **68 yd**
7. **12 in.**   8. **26 ft**

Part B

1. **Rectangle; 72 ft** 2. **Square; 28 yd**
3. **Triangle; 49 in.** 4. **Rectangle; 144 ft**
5. **Triangle; 36 ft**   6. **Rectangle; 86 in.**
7. **Triangle; 44 in.** 8. **Triangle; 27 in.**

Word Problem

Harold needs **88 ft** of fencing.
(32 ft + 12 ft + 32 ft + 12 ft = 88 ft)

EXERCISE 15b (PAGE 98)

1. Area = length × width
   Area = 3 in. × 2 in.
   Area = **6 sq in.**
2. Area = length × width
   Area = 9 ft × 5 ft
   Area = **45 sq ft**
3. Area = length × width
   Area = 8 yd × 8 yd
   Area = **64 sq yd**
4. Area = length × width
   Area = 12 in. × 3 in.
   Area = **36 sq in.**
5. Area = length × width
   Area = 12 ft × 8 ft
   Area = **96 sq ft**
6. Area = length × width
   Area = 16 yd × 16 yd
   Area = **256 sq yd**
7. Area = length × width
   Area = 26 ft × 26 ft
   Area = **676 sq ft**
8. Area = length × width
   Area = 9 in. × 9 in.
   Area = **81 sq in.**

Word Problem

Linda will need to buy **120 tiles**, one tile
for each square foot.
Area = length × width
Area = 12 ft × 10 ft
Area = **120 sq ft**

EXERCISE 15c (PAGE 101)

1. Area = $\dfrac{\text{base} \times \text{height}}{2}$

   Area = $\dfrac{9 \text{ in.} \times 6 \text{ in.}}{2}$

   Area = **27 sq in.**

2. Area = $\dfrac{\text{base} \times \text{height}}{2}$

   Area = $\dfrac{12 \text{ ft} \times 4 \text{ ft}}{2}$

   Area = **24 sq ft**

3. Area = $\dfrac{\text{base} \times \text{height}}{2}$

   Area = $\dfrac{12 \text{ in.} \times 12 \text{ in.}}{2}$

   Area = **72 sq in.**

4. Area = $\dfrac{\text{base} \times \text{height}}{2}$

   Area = $\dfrac{15 \text{ yd} \times 4 \text{ yd}}{2}$

   Area = **30 sq yd**

Word Problem

The area of the counter is **16 sq ft**.

Area = $\dfrac{\text{base} \times \text{height}}{2}$

Area = $\dfrac{8 \text{ ft} \times 4 \text{ ft}}{2}$

Area = 16 sq ft

EXERCISE 16a (PAGE 102)

1.   **Step 1**   **Step 2**   **Step 3**

    1 ft = 12 in.   12   156 in.

                × 13 ft   156 in.

                  36   96 in.

                 12   + 96 in.

               156 in. 504 in.

           Step 4

           42 ft

      12$\overline{)\,504 \text{ in.}}$

          48

          ‾‾

          24

          24

          ‾‾

           0

2.

Step 1	Step 2	Step 3
1 ft = 12 in.	12 × 3 ft ―――― 36 in. 12 × 6 ft ―――― 72 in.	52 in. 36 in. + 72 in. ―――― 160 in.

Step 4

```
     13 ft 4 in.
12) 160 in.
    12
    ――
    40
    36
    ――
     4
```

3.

Step 1	Step 2	Step 3	Step 4
1 yd = 3 ft	3 ×1 yd ―――― 3 ft	3 ft 2 ft +4 ft ―――― 9 ft	3 yd 3) 9 ft 9 ―― 0

4.

Step 1	Step 2	Step 3
1 ft = 12 in.	12 × 2 ft ―――― 24 in.	24 in. 24 in. 36 in. + 36 in. ―――― 120 in.

Step 4

```
     10 ft
12) 120 in.
    12
    ――
    00
```

5.

Step 1	Step 2	Step 3
1 mi = 5280 ft	5280 × 1 mi ―――――― 5280 ft	5280 ft 5280 ft 2000 ft + 2000 ft ―――――― 14560 ft

Step 4

```
           2 mi 4000 ft
5280) 14,560 ft
      10 560
      ――――
       4 000
```

Word Problem

The perimeter of Thelma's garden is **44 ft 8 in.**

Step 1	Step 2
1 ft = 12 in.	14 ft × 12 ―――― 168 in.

Step 3   Step 4

```
168 in.         44 ft 8 in.
168 in.   12) 536 in.
100 in.       48
100 in.       ――
――――          56
536 in.       48
              ――
               8
```

EXERCISE 16b (PAGE 104)

1.

Step 1	Step 2
Area = length × width	Area = length × width
Area = 18 ft × 24 ft	Area = 16 ft × 12 ft
Area = 432 sq ft	Area = 192 sq ft

Step 3

```
  432 sq ft
+ 192 sq ft
――――――
  624 sq ft
```

2.

Step 1	Step 2
Area = length × width	Area = length × width
Area = 22 ft × 12 ft	Area = 14 ft × 14 ft
Area = 264 sq ft	Area = 196 sq ft

Step 3

```
  264 sq ft
+ 196 sq ft
――――――
  460 sq ft
```

3.

Step 1	Step 2
Area = length × width	Area = length × width
Area = 8 ft × 11 ft	Area = 3 ft × 6 ft
Area = 88 sq ft	Area = 18 sq ft

Step 3

```
  88 sq ft
+ 18 sq ft
――――――
 106 sq ft
```

4.

Step 1	Step 2
Area = length × width	Area = length × width
Area = 48 in. × 12 in.	Area = 12 in. × 12 in.
Area = 576 sq in.	Area = 144 sq in.

Step 3

```
  576 sq in.
+ 144 sq in.
――――――
  720 sq in.
```

5.

Step 1	Step 2
Area = length × width	Area = length × width
Area = 32 ft × 8 ft	Area = 5 ft × 3 ft
Area = 256 sq ft	Area = 15 sq ft

Step 3

```
  256 sq ft
+  15 sq ft
――――――
  271 sq ft
```

Word Problem

The area of Pat's L-shaped metal shop is **190 sq ft.**

Step 1

Area = length × width
Area = 10 ft × 14 ft
Area = 140 sq ft

Step 2

Area = length × width
Area = 10 ft × 5 ft
Area = 50 sq ft

Step 3

140 sq ft
+ 50 sq ft
190 sq ft

EXERCISE 16c (PAGE 107)

1. **Step 1**

Area = length × width
Area = 36 in. × 36 in.
Area = 1296 sq in.

Step 2

Area = length × width
Area = 24 in. × 24 in.
Area = 576 sq in.

Step 3

1296 sq in.
− 576 sq in.
720 sq in.

2. **Step 1**

Area = length × width
Area = 30 ft × 12 ft
Area = 360 sq ft

Step 2

Area = length × width
Area = 24 ft × 8 ft
Area = 192 sq ft

Step 3

360 sq ft
− 192 sq ft
168 sq ft

3. **Step 1**

Area = length × width
Area = 2 ft × 2 ft
Area = 4 sq ft

Step 2

Area = length × width
Area = 1 ft × 1 ft
Area = 1 sq ft

Step 3

4 sq ft
− 1 sq ft
3 sq ft

4. **Step 1**

$$Area = \frac{base \times height}{2}$$

$$Area = \frac{8 \text{ ft} \times 6 \text{ ft}}{2}$$

Area = 24 sq ft

Step 2

$$Area = \frac{base \times height}{2}$$

$$Area = \frac{4 \text{ ft} \times 3 \text{ ft}}{2}$$

Area = 6 sq ft

Step 3

24 sq ft
− 6 sq ft
18 sq ft

5. **Step 1**

Area = length × width
Area = 5 yd × 3 yd
Area = 15 sq yd

Step 2

Area = length × width
Area = 3 yd × 2 yd
Area = 6 sq yd

Step 3

15 sq yd
− 6 sq yd
9 sq yd

Word Problem

The area of the sidewalk is **1424 sq yd.**

Step 1

Area = length × width
Area = 180 yd × 180 yd
Area = 32,400 sq yd

Step 2

Area = length × width
Area = 176 yd × 176 yd
Area = 30,976 sq yd

Step 3

32,400 sq yd
− 30,976 sq yd
1424 sq yd

EXERCISE 17 (PAGE 109)

1. 12 + 15 + 25 + 8 = 60
   60 ÷ 4 = **15**

2. $231 + $321 + $132 = $684
   $684 ÷ 3 = **$228**

3. 123 + 56 + 72 + 96 + 65 + 54 + 45 = 511
   511 ÷ 7 = **73**

4. 41 + 42 + 43 + 44 + 45 + 46 = 261
   261 ÷ 6 = **43 r3**

5. 700 + 650 + 900 + 850 + 930 = 4030
   4030 ÷ 5 = **806**

6. 48 + 52 + 48 + 39 + 57 + 50 = 294
   294 ÷ 6 = **49**

7. 54 + 80 + 40 + 48 + 78 = 300
   300 ÷ 5 = **60**

8. $22 + $30 + $18 + $21 + $34 + $307 = $432
   $432 ÷ 6 = **$72**

9. 24 in. + 120 in. + 228 in. + 144 in. + 84 in. + 174 in. = 774 in.
   774 in. ÷ 6 = **129 in.**

10. 236 mi + 52 mi + 18 mi + 558 mi + 241 mi + 1000 mi + 14 mi + 41 mi + 650 mi + 650 mi = 3460 mi
    3460 mi ÷ 10 = **346 mi**

11. 311 + 642 + 213 + 345 + 744 = 2255
    2255 ÷ 5 = **451**

12. 4 + 6 + 24 + 25 + 27 + 34 = 120
    120 ÷ 6 = **20**

13. $9786 + 2435 + 9001 + 3142 = 24{,}364$
    $24{,}364 \div 4 = \textbf{6091}$
14. $90 + 7 + 1 + 34 + 18 + 19 + 33 + 46$
    $= 248$
    $248 \div 8 = \textbf{31}$
15. $\$14 + \$64 + \$72 + \$87 + \$108 = \$345$
    $\$345 \div 5 = \textbf{\$69}$
16. $2\,oz + 4\,oz + 8\,oz + 16\,oz + 32\,oz$
    $+ 64\,oz = 126\,oz$
    $126\,oz \div 6 = \textbf{21 oz}$

Word Problem

Raul's average paycheck per week was **$315**.

$\$325 + \$295 + \$330 + \$350 + \$275$
$= \$1575$
$\$1575 \div 5 = \$315$

EXERCISE 18a (PAGE 111)

1. **(3)** 2. **(4)** 3. **(2)** 4. **(4)** 5. **(1)**

EXERCISE 18b (PAGE 112)

1. **(4)** 2. **(1)** 3. **(5)** 4. **(3)** 5. **(4)**

EXERCISE 18c (PAGE 114)

1. **(4)** 2. **(1)** 3. **(3)** 4. **(2)** 5. **(3)**

EXERCISE 18d (PAGE 116)

There are different ways to estimate solutions. For each problem, one way is shown. You may have estimated in a different way.

1. Actual problem: $\$75 \div 5 =$
   Estimated solution: $\$80 \div 5 = \textbf{\$16}$
2. Actual problem: $7 + 4 + 3 + 7 =$
   Estimated solution: $10 + 5 + 5 + 10 = \textbf{30}$
3. Actual problem: $90 \times 7 =$
   Estimated solution: $100 \times 7 = \textbf{700}$
4. Actual problem: $540 - 395 =$
   Estimated solution: $550 - 400 = \textbf{150}$
5. Actual problem: $56 \times 25 =$
   Estimated solution: $60 \times 25 = \textbf{1500}$

EXERCISE 18e (PAGE 116)

Part A

1. Each person has to pay **$15**.
   ($\$75 \div 5 = \$15$)
2. After the fourth stop, there were **21 people** on the bus. ($7 + 4 + 3 + 7 = 21$)
3. Mario operates his machine **630 times** in a day. ($90 \times 7 = 630$)

4. In December **145 more people** worked at the post office. ($540 - 395 = 145$)
5. In one day **1400 bushels** are picked. ($56 \times 25 = 1400$)

Part B

1. This year **5011 more cars** came off the assembly line. ($31{,}500 - 26{,}489 = 5011$)
2. Jean drives **72 mi each day**.
   ($12\text{ miles} \times 6 = 72\text{ miles}$)
3. There were **25 people** in each car.
   ($375 \div 15 = 25$)
4. Carmine will earn **$6750**.
   ($\$375 \times 18 = \$6750$)
5. At Saturday night's concert there were **4421 people**. ($3576 + 845 = 4421$)
6. **2093 cows** graze in each pasture.
   ($6279 \div 3 = 2093$)
7. The total income for the family is **$40,120**.
   ($\$22{,}320 + \$17{,}500 + \$240 + \$60$
   $= \$40{,}120$)
8. He will have paid **$240** in fees.
   ($\$12 \times 20 = \$240$)
9. Joe will save **$6**. ($\$43 - \$37 = \$6$)
10. There are **4049 more people** living in the town now.
    ($9457 - 5408 = 4049$)

EXERCISE 19 (PAGE 118)

Part A

1. **(3)** 2. **(2)** 3. **(4)** 4. **(3)** 5. **(5)** 6. **(1)**

Part B

1. **(4)** $3\,hr \times \$5 = \$15$ per da;
   $\$15 \times 5\,da = \$75$
2. **(1)** $\$15 \times 6$ cabinets $= \$90$;
   $\$100 - \$90 = \$10$
3. **(2)** $\$15 \div 5$ pairs $= \$3$ per pair;
   $\$3 \times 3$ pairs $= \$9$
4. **(2)** $300\,mi \div 15\,gal = 20\,mi$ per gal;
   $160\,mi \div 20\,mi$ per gal $= 8\,gal$
5. **(4)** $8\,ft \times 2 = 16\,ft$; $16\,ft + 8\,ft = 24\,ft$
6. **(4)** $\$35 - \$7 = \$28$; $\$28 \times 2 = \$56$

WHOLE NUMBERS REVIEW (page 121)

Part A

1. **79**	2. **334**	3. **126**
4. **9**	5. **5601**	6. **19**
7. **160**	8. **95**	9. **32**
10. **915**	11. **108**	12. **119**

13. **132**	14. **19**	15. **2301**
16. **74**	17. **58**	18. **656,420**
19. **26 r26**	20. **868**	21. **40,294**
22. **14**	23. **72,150**	24. **861**
25. **3899**	26. **500**	27. **31,789**
28. **3,135,702**	29. **175**	
30. **7606**	31. **33 r180**	
32. **21,600,000**	33. **645,263**	

Part B

1. **22,221**

2. **Four hundred fifteen thousand, six hundred thirty-two**

3. **370,000**

4. **4**

5. **3000**

6. **12,434**

7. **321**

8. **63**

9. **3 ft**

10. **7157**

11. **10,764**

12. **1008 sq ft**

GED PRACTICE 1 (page 122)

1. **(4)** 87 + 302 + 234 + 106 = 729
2. **(4)** 3894 + 766 = 4660
3. **(1)** $19,034 − $17,250 = $1784
4. **(2)** 2532 − 460 = 2072
5. **(2)** 1211 mi × 12 = 14,532 mi
6. **(2)** $4640 × 809 = $3,753,760
7. **(3)** $500,000 × 4 = $125,000
8. **(1)** 216 mi ÷ 18 = 12 mi

9. **(2)**

```
   6 ft 10 in.
   6 ft  2 in.          2 ft
   6 ft  1 in.      12) 24 in.
   5 ft  9 in.
 + 5 ft  2 in.
  28 ft 24 in.
```

```
  28 ft          6 ft
 + 2 ft       5) 30 ft
  30 ft
```

10. **(4)**

```
   1 ft 2 in.
 ×      4
   4 ft 8 in.
```

11. **(2)**

```
    40,963                40,565
    37,461            5 ) 202,825
    40,066               20
    34,312                2
 +  50,023                0
   202,825                2 8
                          2 5
                           32
                           30
                           25
                           25
                            0
```

12. **(3)**

Area = length × width
Area = 2 ft × 4 ft
Area = 8 sq ft

Area = length × width   8 sq ft
Area = 3 ft × 2 ft   + 6 sq ft
Area = 6 sq ft   14 sq ft

13. **(5)**

```
 $ 24          $150
   49         − 118
   33          $ 32
    8
 +  4
 $118
```

14. **(4)**

```
     40 lb          18
 4) 160 lb       × 40 lb
    16            720 lb
    00
```

UNIT 2   DECIMALS

EXERCISE 20a (PAGE 128)

Part A

1. (a) **.51** (b) **.125** (c) **.457** (d) **.2379** (e) **.75**
2. (a) **.837** (b) **.73** (c) **.3784** (d) **.783** (e) **.842**
3. (a) **.3492** (b) **.99432** (c) **.344229** (d) **.92939** (e) **.42234**

Part B

1. .52 has **5** tenths and **2** hundredths.
2. .265 has **2** tenths, **6** hundredths, and **5** thousandths.
3. .8925 has **8** tenths, **9** hundredths, **2** thousandths, and **5** ten thousandths.
4. .98523 has **9** tenths, **8** hundredths, **5** thousandths, **2** ten thousandths, and **3** hundred thousandths.
5. .673127 has **6** tenths, **7** hundredths, **3** thousandths, **1** ten thousandth, **2** hundred thousandths, and **7** millionths.

EXERCISE 20b (PAGE 129)

1. In .08, the 8 is in the **hundredths** place.
2. In .081, the 0 holds the **tenths** place.
3. In .002003, the 2 is in the **thousandths** place and the 3 is in the **millionths** place.
4. .017 has **0** tenths, **1** hundredth, and **7** thousandths.
5. In .09, the digit in the tenths place is a **0**.

EXERCISE 21a (PAGE 129)

1. **forty-two hundredths**
2. **thirty-five hundredths**
3. **eighty-two hundredths**
4. **seventy-one hundredths**
5. **ninety-eight hundredths**
6. **four tenths**
7. **six hundred thirteen thousandths**
8. **three hundredths**
9. **seven hundred sixty-three thousandths**
10. **three thousandths**
11. **two thousand, one hundred twelve ten thousandths**
12. **two thousand, one hundred four ten thousandths**
13. **nine hundred nine thousandths**
14. **two thousand, five hundred seventy-one ten thousandths**
15. **seven ten thousandths**
16. **three thousand, two hundred forty-five ten thousandths**
17. **twelve hundredths**
18. **four hundred twelve thousandths**
19. **seventy-eight thousandths**
20. **fifty-three thousand, two hundred eight hundred thousandths**
21. **four tenths**
22. **twenty-two thousand, three hundred sixty-four hundred thousandths**
23. **fourteen millionths**
24. **three thousand, one hundred sixty-seven ten thousandths**
25. **six millionths**

EXERCISE 21b (PAGE 130)

Part A

1. **(c)**
2. **(d)**
3. **(b)**
4. **(e)**
5. **(a)**
6. **(h)**
7. **(f)**
8. **(i)**
9. **(g)**
10. **(j)**

Part B

1. **thirty-six hundredths**
2. **nine tenths**
3. **four hundredths**
4. **fifty-two hundredths**
5. **thirty-seven hundredths**
6. **four hundred seventy-eight thousandths**
7. **twenty-one ten thousandths**
8. **sixteen and four hundred seventy-eight thousandths**
9. **two and six tenths**
10. **twenty-one and four hundredths**

Word Problem

Expressed in words, the length of the piece is **fifteen hundredths of an inch.**

EXERCISE 21c (PAGE 132)

1. **.06**	6. **1.009**	11. **.003**
2. **.007**	7. **2.0036**	12. **38.0803**
3. **.0008**	8. **20.02**	13. **150.015**
4. **.00048**	9. **803.083**	14. **80.08**
5. **.036**	10. **77.77**	15. **6050.056**

Word Problem

The pump showed **3.3 gal** after Dave gave Rudy the amount of gas he asked for.

EXERCISE 22a (PAGE 133)

Part A

1. **.6**	2. **.55**	3. **.1**
4. **.76**	5. **.90**	6. **.320**
7. **.73**	8. **.48**	9. **.15**
10. **.180**	11. **.401**	12. **.6**
13. **5.25**	14. **17.5**	15. **89.8**
16. **4.5**		

Part B

1. **.04**	2. **.12**	3. **.3**
4. **.59**	5. **.312**	6. **.489**
7. **.5**	8. **.1**	9. **.125**
10. **.22**	11. **.312**	12. **.07**
13. **1.066**	14. **4.05**	15. **3.04**
16. **20.7**		

Word Problem

The thicker sheet was the one **.025 in.** thick.

EXERCISE 22b (PAGE 135)

Part A

1. 4.01, 3.34, 3.28
2. 8.4, 7.41, 7.34
3. 5.91, 5.9, 5.87
4. 3.96, 2.8, 2.69
5. 7.35, 5.73, 5.37
6. 2.55, 2.3, 2.15
7. 3.088, 3.08, 3.07
8. .693, .655, .6

Part B

1. 1.7, 2.5, 2.7, 3.9
2. 2.43, 3.4, 4.13, 4.23
3. .6, .821, 1.02, 1.68
4. 34.07, 34.7, 40.3, 43.7
5. .99, 9.09, 9.9, 9.99
6. 4.889, 4.89, 4.980, 4.983
7. .60, .604, .64, .641
8. .229, .23, .234, .28

Word Problem

The most rain (3.14 in.) fell on **the fourth day**. The least rain (.87 in.) fell on **the second day**.

EXERCISE 23a (PAGE 136)

Part A

1. 4.5   2. 15.3   3. 48.1   4. 224.6
5. 96.3   6. .3   7. .4   8. .4
9. .8   10. .4

Part B

1. .64   2. .67   3. .85   4. .40
5. .44   6. 4.26   7. 9.14   8. 14.06
9. 68.35   10. 156.05

Part C

1. .270   2. .678   3. 2.430   4. 9.895
5. 82.601   6. 1.886   7. 23.224
8. 8.312   9. 76.005   10. 11.235

Word Problem

Jason told his friends the fish weighed **7.5 lb.**

EXERCISE 23b (PAGE 138)

1. 6   2. 15   3. 49   4. 13
5. 7   6. 12   7. 5   8. 32
9. 17   10. 2   11. $6   12. $4
13. $77   14. $11   15. $4   16. 8
17. 24   18. 1265   19. 14   20. 57

Word Problem

Rounded to the nearest dollar, Jeff has $6 in his pocket.

EXERCISE 24a (PAGE 140)

1. 3.5   2. .9   3. $.78
4. 8.88   5. $.99   6. 2.108
7. 3.164   8. .86   9. $49
10. 6.532   11. 9.313   12. 9.96
13. .42   14. .691   15. 5.83

Word Problem

The steaks weighed **.79 lb** altogether.
(.37 + .42 = .79)

EXERCISE 24b (PAGE 141)

1. 1.2   2. 1.3   3. .91
4. $1.71   5. .91   6. 1.02
7. 1.11   8. .38   9. .63
10. .912   11. .69   12. .83
13. .92   14. 1.27   15. .75
16. 4.63   17. 91.83   18. 136.36
19. 6.35   20. 12.26   21. 17.18
22. 4.506   23. 8.131   24. 30.196
25. $43.44   26. $8.97   27. 17.81
28. 1.031   29. .81   30. .492
31. .8   32. 1.009   33. .9817
34. 1.5131   35. .7432   36. 1.03
37. 658.303   38. 13.64
39. 1060.276   40. .73

Word Problem

In all, Salinda spent **$2.75** those 3 days.
($.70 + $1.20 + $.85 = $2.75)

EXERCISE 24c (PAGE 142)

1. $48.23   2. 10.072
3. 100.05   4. 308.03
5. 59.94   6. 130.01
7. 46.12   8. 69.86
9. 396.003   10. 873.631
11. 98.1   12. 110.07
13. 88.348   14. 99.77
15. 43.9

Word Problem

Miranda spent **$290.94** in all.
($27.99 + $195 + $67.95 = $290.94)

EXERCISE 25a (PAGE 144)

1. .4   2. .3   3. $.31
4. .53   5. $.52   6. .11
7. $.22   8. .401   9. .203
10. .322   11. $.38   12. .17
13. .245   14. .0505   15. .2224

16. .34   17. .059   18. .0018
19. .183   20. .226   21. $64.02
22. 73.216   23. 11.4   24. 29.12
25. 5.832   26. 35.21   27. 44.101
28. 3   29. 1.1   30. .1001
31. 21.43   32. 11.03   33. $11.11
34. 34.11   35. 12.4   36. 111.002
37. .22   38. 22.24
39. 200.07   40. 13.222

Word Problem

The sale price of the sweaters was **$33.45**.
($39.95 − $6.50 = $34.45)

EXERCISE 25b (PAGE 145)

1. **2.9**   2. **9.9**   3. **310.9**
4. **$4.56**   5. **1.95**   6. **22.19**
7. **$37.99**   8. **17.597**   9. **8.671**
10. **74.987**   11. **4.5804**   12. **344.06**
13. **1.6036**   14. **207.67**   15. **3.688**
16. **29.37**   17. **28.89**   18. **6.9**
19. **$1.88**   20. **$18**

Word Problem

The lighter chip weighs **.193 oz** less.
(.243 oz − .05 oz = .193 oz)

EXERCISE 25c (PAGE 146)

1. **83.38**   2. **74.95**   3. **67.32**
4. **22.87**   5. **10.97**   6. **95.11**
7. **41.69**   8. **63.54**   9. **87.933**
10. **35.446**   11. **131.997**   12. **311.958**
13. **655.222**   14. **110.5**   15. **230.911**
16. **622.377**   17. **474.73**   18. **7.6944**
19. **11.6896**   20. **343.567**

Word Problem

The difference between the regular price
and the sale price is **$12.05**.
($42 - $29.95 = $12.05)

MIXED PRACTICE 3 (page 148)

1. **.798**   2. **.34**   3. **.228**
4. **17.66**   5. **7.5**   6. **4.154**
7. **11.83**   8. **.81**   9. **.335**
10. **31.68**   11. **57.4**   12. **41.41**
13. **$4.06**   14. **19.6**   15. **5.7**
16. **$3.17**   17. **45.8**   18. **5.798**
19. **6.344**   20. **$1.61**   21. **.48**
22. **52.872**   23. **.37**   24. **.427**
25. **6.96**   26. **5.2**   27. **3.51**

28. **178.06**   29. **159.948**   30. **6.19**
31. **$3.02**   32. **$17.50**   33. **$11.31**
34. **2.9**   35. **8.25**   36. **.14**
37. **.113**   38. **14.81**   39. **28.006**
40. **$59.35**   41. **9.35**
42. **93.182**   43. **$1.45**
44. **$17.32**   45. **62.12**
46. **76.978**   47. **13.646**
48. **3.737**   49. **$20.02**
50. **$21.98**

EXERCISE 26a (PAGE 149)

Part A

1. **1.00**   2. **2.88**   3. **.33**
4. **19.44**   5. **48.633**   6. **3.63**
7. **4.946**   8. **4.73**   9. **3.08**
10. **1.98**

Part B

1. **32.504**   2. **$.68**   3. **$2.08**
4. **7.2912**   5. **27.009**   6. **18.66**
7. **$.72**   8. **$2.05**   9. **.852**
10. **8.084**   11. **$1.70**   12. **2.090**
13. **4.0518**   14. **$8.32**   15. **149.6**
16. **10.625**   17. **$31.88**   18. **11.625**
19. **2.5725**   20. **7.256**   21. **1.04**
22. **46.93**   23. **7.276**   24. **25.11**
25. **3.20**   26. **$13.25**   27. **16.641**
28. **54.3616**   29. **3.29**   30. **$54.72**

Word Problem

All 4 packages weigh **3.56 lb**.
(.89 lb × 4 = 3.56 lb)

EXERCISE 26b (PAGE 151)

Part A

1. **.40448**   2. **.18**   3. **.14175**
4. **.819**   5. **.285396**

Part B

1. **.11024**   2. **.168**   3. **.10044**
4. **.455**   5. **$.22**   6. **.161**
7. **.731224**   8. **$.65**   9. **.264**
10. **$.13**   11. **.78016**   12. **.1505798**

Word Problem

Mr. Tang paid **$.44** for the potatoes.
($.49 × .9 = $.441, so the answer is
rounded to the nearest cent.)

Part A

1. **4.0448**   2. **.48**   3. **141.75**
4. **8.19**   5. **285.396**

Part B

1. **.1488**   2. **$.39**   3. **1.608**
4. **25.839**   5. **119.328**   6. **$.79**
7. **7.26**   8. **5.5854**   9. **$.05**
10. **8.217**

Word Problem

Nina makes **$241.13** a week.
($6.43 × 37.5 = $241.125, so the
answer is rounded to the nearest cent.)

EXERCISE 26d (PAGE 154)

Part A

1. **.0414**   2. **.072820**   3. **.002367**
4. **.0847**   5. **.0209**

Part B

1. **.06**   2. **$.09**   3. **.0066**
4. **.00924**   5. **.08255**   6. **.00159**
7. **.000344**   8. **.00027**   9. **.004**
10. **.0000207**   11. **$.07**   12. **.01556**
13. **.0171**   14. **.001848**   15. **.003292**
16. **$.00**

Word Problem

Lavatria paid **$.07** for the bones.
($.35 × .2 = $.07)

EXERCISE 26e (PAGE 155)

1. **410**   2. **3700**   3. **74**
4. **400**   5. **65.55**   6. **780**
7. **25**   8. **310**   9. **6472**
10. **56.3**   11. **7800**   12. **7.4**
13. **305**   14. **42,123**   15. **67.8**

Word Problem

Laura walks **12.5 mi** in a week between
her home and her job.
(1.25 mi × 10 = 12.5 mi)

EXERCISE 27a (PAGE 156)

1. **.4**   2. **.3**   3. **$.11**
4. **.21**   5. **$.31**   6. **1.3**
7. **2.1**   8. **.91**   9. **$.81**
10. **.62**   11. **.007**   12. **.006**

13. **.004**   14. **.007**   15. **.003**
16. **$.61**   17. **.342**   18. **.07**
19. **.82**   20. **6.7**

Word Problem

Each person paid **$3.19**.
($6.38 ÷ 2 = $3.19)

EXERCISE 27b (PAGE 157)

1. **.08**   2. **.05**   3. **.04**   4. **.06**
5. **.05**   6. **.05**   7. **.05**   8. **.05**
9. **.005**   10. **.04**   11. **.05**   12. **.05**
13. **.04**   14. **.059**   15. **.009**   16. **.003**

Word Problem

In each part, **.075 ton** should be shipped.
(.6 ton ÷ 8 = .075 ton)

EXERCISE 27c (PAGE 158)

1. **.35**   2. **.076**   3. **.65**   4. **.085**
5. **2.05**   6. **.62**   7. **.0125**   8. **.0375**
9. **.25**   10. **.12**   11. **.2345**   12. **.00625**
13. **.975**   14. **.16**   15. **.175**   16. **.032**

Word Problem

Each portion will weigh **1.125 lb**.
(4.5 lb ÷ 4 = 1.125 lb)

EXERCISE 27d (PAGE 160)

Part A

1. **.5**   2. **.3**   3. **.0**   4. **.6**   5. **.0**

Part B

1. **.19**   2. **.07**   3. **.32**   4. **.12**   5. **.30**

Part C

1. **1.157**   2. **8.217**   3. **.417**   4. **3.833**
5. **.314**

Word Problem

Each employee contributed **$8.71** to the
cost of the coffee maker.
($60.95 ÷ 7 = $8.707, which, rounded
to the nearest cent, is $8.71)

EXERCISE 27e (PAGE 161)

1. **8**   2. **7**   3. **.9**
4. **8**   5. **4**   6. **.58**
7. **.53**   8. **145**   9. **6**
10. **35.6**   11. **30**   12. **5**
13. **732**   14. **14**   15. **300**
16. **3**   17. **22.45**   18. **250**
19. **15**   20. **35**

Word Problem

Luz packed **23 boxes**.
(57.5 lb ÷ 2.5 = 23 boxes)

EXERCISE 27f (PAGE 162)

1. **80**	2. **50**	3. **90**
4. **50**	5. **70**	6. **30**
7. **2**	8. **300**	9. **5000**
10. **5000**	11. **80**	12. **230**
13. **800**	14. **8000**	15. **8000**
16. **600**	17. **7400**	18. **700**
19. **5,280,000**	20. **1400**	

Word Problem

Harry packed **600 packages**.
(480 lb ÷ .8 lb = 600 packages)

EXERCISE 27g (PAGE 163)

1. **.032**	2. **.0045**	3. **.06**
4. **.009**	5. **.00004**	6. **.089**
7. **.0036**	8. **.132**	9. **.0009**
10. **.012**	11. **2.64**	12. **.0118**
13. **.6816**	14. **.0067**	15. **.000834**
16. **.023**		

Word Problem

Each piece was **1.64 ft** long.
(16.4 ft ÷ 10 = 1.64 ft)

MIXED PRACTICE 4 (page 165)

1. **$1.80**	2. **.2262**	3. **.28**
4. **.240**	5. **16.2**	6. **.03**
7. **300**	8. **1.3**	9. **.00847**
10. **2.7**	11. **10.62**	12. **21.35**
13. **.19672**	14. **130.36**	15. **392.42**
16. **8**	17. **23**	18. **15**
19. **120**	20. **69**	21. **9.275**
22. **31.35**	23. **1015.2**	24. **$71.20**
25. **$102.86**	26. **6**	27. **2.5**
28. **.015**	29. **5**	30. **8**
31. **.29388**	32. **313.5**	33. **$14.75**
34. **123.6206**	35. **.195**	36. **.0724**
37. **1590**	38. **40**	39. **27.8**
40. **8.36**	41. **$493.64**	42. **.0792**
43. **$2664.18**	44. **269.28**	45. **.0378**
46. **$.04**	47. **$1.25**	48. **8.02**
49. **41.07**	50. **15**	

DECIMAL SKILLS REVIEW (page 166)

Part A

1. **3.987**	2. **$100**	3. **2.046**
4. **2.11**	5. **1.977**	6. **8.111**
7. **$43.50**	8. **.0301**	9. **6.85**
10. **2.11**	11. **.2871**	12. **$94.85**
13. **1.026**	14. **$.23**	15. **2.35**
16. **.403**	17. **374.9**	18. **594.128**
19. **.14848**	20. **.05**	21. **7.2**
22. **30.989**	23. **708.615**	24. **.005**
25. **19.85**	26. **$45.43**	27. **.0384**
28. **15.999**	29. **21.276**	30. **.2345**
31. **3.833**	32. **.0954**	33. **65.3**
34. **3.1**	35. **24**	36. **1100**
37. **600**	38. **160**	39. **2000**
40. **12.21**		

Part B. Solve each problem.

1. The gas pump showed **10.12 gal.**
2. **The smoked fish** weighed the most, and **the sliced ham** weighed the least.
3. Listed in order of size from smallest to largest, the 5 items are **(e)**, **(b)**, **(a)**, **(c)**, and **(d)**.
4. George should have reported the following: (a) **7.1 lb**, (b) **25.7 km**, (c) **18.0 g**, and (d) **.1 cm.**
5. The total rainfall for the four days was **3.3 in.**
(1.09 in. + .88 in. + 1.13 in. + .2 in. = 3.3 in.)
6. The heavier part weighs **.034 g** more than the lighter one.
(.43 g − .396 g = .034 g)
7. The total cost is $**.06.**
(.8 oz × $.07 = $.056)
8. Each seed weighs **.019 oz.**
(.171 oz ÷ 9 = .019 oz)

EXERCISE 28 (PAGE 167)

1. Rita earned **$96.30** in all.
($5.35 × 18 = $96.30)
2. Ted Williams's batting average was **.063** point more. (.406 − .343 = .063)
3. Marge has opened the wrench **.8 in.** altogether. (.6 in. + .2 in. = .8 in.)
4. **288 boxes** of chocolate are packed.
(360 lb ÷ 1.25 = 288 boxes)
5. Leo jogs **22.75 mi** in one week.
(3.25 mi × 7 = 22.75 mi)

6. Luis took **1.28 sec** less to drive around the track.
   (45.04 sec − 43.76 sec = 1.28 sec)

7. The total amount of the checks Alicia wrote was **$513.65**.
   ($300 + $45.76 + $167.89 = $513.65)

8. The total distance Carlos drives between home and work each week is **356 mi**.
   (35.6 mi × 10 = 356 mi)

9. There are **40** sheets in the stack.
   (60 in. ÷ 1.5 in. = 40)

10. All the bolts weighed **807 ounces**.
    (1076 × .75 oz = 807 oz)

EXERCISE 29a (PAGE 169)

1. A milliliter is **.001 liter**.
2. A hectogram is **100 grams**.
3. A centimeter is **.01 meter**.
4. A deciliter is **.1 liter**.
5. A dekagram is **10 grams**.

EXERCISE 29b (PAGE 170)

1. 2.5 kg = **2500 g**   2. 1.2 daL = **12 L**
3. 2 hm = **200 m**   4. 200 dm = **20 m**
5. 40 mm = **4 cm**   6. .2 cm = **.002 m**
7. 4.5 L = **4500 mL**
8. 5.03 km = **5,030,000 mm**
9. 960 g = **.96 kg**   10. 344.8 L = **.3448 kL**

Word Problem

2.5 k of raisins is the same as **2500 g** of raisins.

EXERCISE 30 (PAGE 171)

1. 1.2 + 1.5 + 2.5 + 8 = 13.2
   13.2 ÷ 4 = **3.3**

2. $2.31 + $3.21 + $1.32 = $6.84
   $6.84 ÷ 3 = **$2.28**

3. .123 + 5.8 + .72 + 96 + 6.5 + 54 + .45 = 163.593
   163.593 ÷ 7 = **23.370**

4. 4.1 + .42 + .043 + .044 + .145 + 4.06 = 8.812
   8.812 ÷ 6 = **1.469**

5. 700.65 + 900.85 + 930 = 2531.50
   2531.50 ÷ 3 = **843.833**

6. .48 + .52 + .48 + .39 + .57 + .5 = 2.94
   2.94 ÷ 6 = **.49**

7. 2.54 + 3.8 + 9.4 + 4.8 + 7.815 = 28.355
   28.355 ÷ 5 = **5.671**

8. $.22 + $.30 + $.18 + $.21 + $.34 + $3.09 = $4.34
   $4.34 ÷ 6 = **$.72**

9. 2.4 m + 1.2 m + 2.28 m + 1.44 m + 8.4 m + 17.7 m = 33.42 m
   33.42 m ÷ 6 = **5.57 m**

10. 2.3 km + 5.2 km + 1.8 km + 5.5 km + 2.4 km + 10 km + .4 km + .4 km + 6.5 km + 6 km = 40.5 km
    40.5 km ÷ 10 = **4.05 km**

Word Problem

The average of the readings was **100.0**.
   99.2 + 100.6 + 101.1 + 99.8 + 99.1 = 499.8
   499.8 ÷ 5 = 99.96, which is 100.0 rounded to the nearest tenth.

EXERCISE 31a (PAGE 173)

1. **17.28 cm**   2. **12.8 m**   3. **20.8 km**

Word Problem

To enclose the garden, **15 m** of fencing are needed.
   (3.5 m + 4 m + 3.5 m + 4 m = 15 m)

EXERCISE 31b (PAGE 175)

1. Area = length × width
   Area = 8.1 dam × 8.1 dam
   Area = **65.61 sq dam**

2. Area = $\dfrac{\text{base} \times \text{height}}{2}$

   Area = $\dfrac{4.4 \text{ hm} \times 2.2 \text{ hm}}{2}$

   Area = **4.84 sq hm**

3. Area = length × width
   Area = 17.4 cm × 12.1 cm
   Area = **210.54 sq cm**

Word Problem

To cover the floor, **14 sq m** of linoleum are needed. (3.5 m × 4 m = 14 sq m)

EXERCISE 31c (PAGE 175)

1. Perimeter = **26.6 km**

   Area = $\dfrac{\text{base} \times \text{height}}{2}$

   Area = $\dfrac{12.6 \text{ km} \times 3.05 \text{ km}}{2}$

   Area = **19.215 sq km**

2. Perimeter = **21.6 cm**

   Area = length × width
   Area = 5.4 cm × 5.4 cm
   Area = **29.16 sq cm**

3. Perimeter = **13.8 mm**

   Area = $\dfrac{\text{base} \times \text{height}}{2}$

   Area = $\dfrac{4.6 \text{ mm} \times 3.99 \text{ mm}}{2}$

   Area = **9.177 sq mm**

4. Perimeter = **16.8 dm**

   Area = length × width
   Area = 4.2 dm × 4.2 dm
   Area = **17.64 sq dm**

5. Perimeter = **44.6 km**

   Area = length × width
   Area = 8 km × 14.3 km
   Area = **114.4 sq km**

6. Perimeter = **36.4 hm**

   Area = length × width
   Area = 6.5 hm × 11.7 hm
   Area = **76.05 sq hm**

7. Perimeter = **13.2 hm**

   Area = $\dfrac{\text{base} \times \text{height}}{2}$

   Area = $\dfrac{4.4 \text{ hm} \times 3.8 \text{ hm}}{2}$

   Area = **8.36 sq hm**

8. Perimeter = **50.4 dam**

   Area = length × width
   Area = 12.6 dam × 12.6 dam
   Area = **158.76 sq dam**

9. Perimeter = **12.8 m**

   Area = length × width
   Area = 4.4 m × 2 m
   Area = **8.8 sq m**

Word Problem

   The perimeter of the garden is **65.8 m.**

   14.3 m + 18.6 m + 14.3 m + 18.6 m
     = 65.8 m

   The area of the garden is **265.98 sq m.**

   14.3 m × 18.6 m = 265.98 sq m

EXERCISE 32a (PAGE 178)

1. diameter = **30 ft**   2. radius = **.625 cm**

3. diameter = **31.5 m**

4. diameter = **400.6 dm**

5. radius = **6.8 cm**   6. diameter = **98 in.**

7. radius = **3.75 km**   8. radius = **33 yd**

9. radius = **9.7 m**   10. diameter = **47 mi**

EXERCISE 32b (PAGE 180)

1. Circumference = 3.14 × diameter
   Circumference = 3.14 × 6 in.
   Circumference = **18.84 in.**

2. Circumference = 3.14 × diameter
   Circumference = 3.14 × 10 ft
   Circumference = **31.4 ft**

3. Circumference = 3.14 × diameter
   Circumference = 3.14 × 1.6 m
   Circumference = **5.024 m**

4. Circumference = 3.14 × diameter
   Circumference = 3.14 × .8 cm
   Circumference = **2.512 cm**

5. Circumference = 3.14 × diameter
   Circumference = 3.14 × 325 ft
   Circumference = **1020.5 ft**

6. Circumference = 3.14 × diameter
   Circumference = 3.14 × 40 yd
   Circumference = **125.6 yd**

7. Circumference = 3.14 × diameter
   Circumference = 3.14 × 64 mi
   Circumference = **200.96 mi**

8. Circumference = 3.14 × diameter
   Circumference = 3.14 × .18 m
   Circumference = **.565 m**

9. Circumference = 3.14 × diameter
   Circumference = 3.14 × 1.1 mm
   Circumference = **3.454 mm**

10. Circumference = 3.14 × diameter
    Circumference = 3.14 × 10.5 km
    Circumference = **32.97 km**

EXERCISE 32c (PAGE 181)

1. diameter = 1.2 m
   Circumference = 3.14 × diameter
   Circumference = 3.14 × 1.2 m
   Circumference = **3.768 m**

2. diameter = 20 ft
   Circumference = 3.14 × diameter
   Circumference = 3.14 × 20 ft
   Circumference = **62.8 ft**

3. diameter = 21 m
   Circumference = 3.14 × diameter
   Circumference = 3.14 × 21 m
   Circumference = **65.94 m**

4. diameter = 1.6 cm
   Circumference = 3.14 × diameter
   Circumference = 3.14 × 1.6 cm
   Circumference = **5.024 cm**

5. diameter = 6.42 dam
   Circumference = 3.14 × diameter
   Circumference = 3.14 × 6.42 dam
   Circumference = **20.159 dam**

6. diameter = 80 yd
   Circumference = 3.14 × diameter
   Circumference = 3.14 × 80 yd
   Circumference = **251.2 yd**

7. diameter = 128 mi
   Circumference = 3.14 × diameter
   Circumference = 3.14 × 128 mi
   Circumference = **401.92 mi**

8. diameter = .36 m
   Circumference = 3.14 × diameter
   Circumference = 3.14 × .36 m
   Circumference = **1.130 m**

9. diameter = 2.2 mm
   Circumference = 3.14 × diameter
   Circumference = 3.14 × 2.2 mm
   Circumference = **6.908 mm**

10. diameter = 3 km
    Circumference = 3.14 × diameter
    Circumference = 3.14 × 3 km
    Circumference = **9.42 km**

EXERCISE 32d (PAGE 182)

1. Area = 3.14 × radius × radius
   Area = 3.14 × 14 in. × 14 in.
   Area = **615.44 sq in.**

2. Area = 3.14 × radius × radius
   Area = 3.14 × 1.5 cm × 1.5 cm
   Area = **7.065 sq cm**

3. Area = 3.14 × radius × radius
   Area = 3.14 × 2.3 mi × 2.3 mi
   Area = **16.611 sq mi**

4. Area = 3.14 × radius × radius
   Area = 3.14 × 1 m × 1 m
   Area = **3.14 sq m**

5. Area = 3.14 × radius × radius
   Area = 3.14 × 44 yd × 44 yd
   Area = **6079.04 sq yd**

6. Area = 3.14 × radius × radius
   Area = 3.14 × .85 km × .85 km
   Area = **2.269 sq km**

7. Area = 3.14 × radius × radius
   Area = 3.14 × 13 in. × 13 in.
   Area = **530.66 sq in.**

8. Area = 3.14 × radius × radius
   Area = 3.14 × .05 m × .05 m
   Area = **.008 sq m**

9. Area = 3.14 × radius × radius
   Area = 3.14 × 1.1 mm × 1.1 mm
   Area = **3.799 sq mm**

10. Area = 3.14 × radius × radius
    Area = 3.14 × 10.5 km × 10.5 km
    Area = **346.185 sq km**

EXERCISE 32e (PAGE 183)

1. radius = 3 in.
   Area = 3.14 × radius × radius
   Area = 3.14 × 3 in. × 3 in.
   Area = **28.26 sq in.**

2. radius = 5 ft
   Area = 3.14 × radius × radius
   Area = 3.14 × 5 ft × 5 ft
   Area = **78.5 sq ft**

3. radius = .8 m
   Area = 3.14 × radius × radius
   Area = 3.14 × .8 m × .8 m
   Area = **2.010 sq m**

4. radius = .4 cm
   Area = 3.14 × radius × radius
   Area = 3.14 × .4 cm × .4 cm
   Area = **.502 sq cm**

5. radius = 162 ft
   Area = 3.14 × radius × radius
   Area = 3.14 × 162 ft × 162 ft
   Area = **82,406.16 sq ft**

6. radius = 20 yd
   Area = 3.14 × radius × radius
   Area = 3.14 × 20 yd × 20 yd
   Area = **1256 sq yd**

7. radius = 32 mi
   Area = 3.14 × radius × radius
   Area = 3.14 × 32 mi × 32 mi
   Area = **3215.36 sq mi**

8. radius = .09 m
   Area = 3.14 × radius × radius
   Area = 3.14 × .09 m × .09 m
   Area = **.025 sq m**

9. radius = .56 mm
   Area = 3.14 × radius × radius
   Area = 3.14 × .56 mm × .56 mm
   Area = **.985 sq mm**

10. radius = 5.25 km
    Area = 3.14 × radius × radius
    Area = 3.14 × 5.25 km × 5.25 km
    Area = **86.546 sq km**

Word Problem
   The radius of Carolyn's garden was **3 yd.**
   The diameter was **6 yd.**
   The circumference was **18.84 yd.**
   (3.14 × 6 yd = 18.84 yd)
   The area was **28.26 sq yd.**
   (3.14 × 3 yd × 3 yd = 28.26 sq yd)

EXERCISE 33a (PAGE 184)

1. Cost = number × rate
   Cost = 5.67 lb × $1.39
   Cost = **$7.88**

2. Cost = number × rate
   Cost = 1.3 lb × $2.49
   Cost = **$3.24**

3. Cost = number × rate
   Cost = 52 ft × $.39
   Cost = **$20.28**

4. Cost = number × rate
   Cost = 4.75 lb × $3.49
   Cost = **$16.58**

5. Cost = number × rate
   Cost = 250 bricks × $.43
   Cost = **$107.50**

6. Cost = number × rate
   Cost = 7.5 yd × $1.30
   Cost = **$9.75**

7. Cost = number × rate
   Cost = 6 pairs × $3.25
   Cost = **$19.50**

8. Cost = number × rate
   Cost = 20 tokens × $1.15
   Cost = **$23.00**

9. Cost = number × rate
   Cost = 1.4 lb × $2.58
   Cost = **$3.61**

10. Cost = number × rate
    Cost = 4 sweaters × $34.75
    Cost = **$139.00**

EXERCISE 33b (PAGE 185)

1. Unit cost = price ÷ unit
   Unit cost = $.99 ÷ 1.2 lb
   Unit cost = **$.83 per lb**

2. Unit cost = price ÷ unit
   Unit cost = $1.79 ÷ .75 lb
   Unit cost = **$2.39 per lb**

3. Unit cost = price ÷ unit
   Unit cost = $.81 ÷ 18 oz
   Unit cost = **$.05 per oz**

4. Unit cost = price ÷ unit
   Unit cost = $1.29 ÷ 6 oz
   Unit cost = **$.22 per oz**

5. Unit cost = price ÷ unit
   Unit cost = $3.28 ÷ 3.6 gal
   Unit cost = **$.91 per gal**

6. Unit cost = price ÷ unit
   Unit cost = $1.45 ÷ 1.5 lb
   Unit cost = **$.97 per lb**

7. Unit cost = price ÷ unit
   Unit cost = $5.75 ÷ 4.66 lb
   Unit cost = **$1.23 per lb**

8. Unit cost = price ÷ unit
   Unit cost = $17.50 ÷ 14 lb
   Unit cost = **$1.25 per lb**

9. Unit cost = price ÷ unit
   Unit cost = $5.60 ÷ 1.12 lb
   Unit cost = **$5.00 per lb**

10. Unit cost = price ÷ unit
    Unit cost = $13.80 ÷ .6 oz
    Unit cost = **$23.00 per oz**

EXERCISE 34a (PAGE 187)

Part A

1. Don's net pay was **$301.98.**
   $60.56 + $29.81 = $90.37 total deductions
   $392.35 − $90.37 = $301.98 net pay

2. Joan's net pay was **$370.18.**
   $77.12 + $36.72 = $113.84 total deductions
   $484.02 − $113.84 = $370.18 net pay

3. Mark's net pay was **$391.40.**
   $82.16 + $39.22 = $121.38 total deductions
   $512.78 − $121.38 = $391.40 net pay

4. Ruben's net pay was **$624.54.**
   $151.20 + $64.26 = $215.46 total deductions
   $840.00 − $215.46 = $624.54 net pay

5. Sam's net pay was **$1085.51.**
   $262.80 + $111.69 = $374.49 total deductions
   $1460.00 − $374.49 = $1085.51 net pay

Part B

1. Sarah's take-home pay was **$299.35.**
   $73.93 + $31.42 + $2.63 + $3.42 = $111.40 total deductions
   $410.75 − $111.40 = $299.35 net pay

2. Vic's net pay was **$272.72.**
   $76.41 + $32.47 + $42.90 = $151.78 total deductions
   $424.50 − $151.78 = $272.72 net pay

3. Kathy's net pay was **$649.76.**
   $173.66 + $67.53 + $73.80 = $314.99 total deductions
   $964.75 − $314.99 = $649.76 net pay

4. Hans's net pay was **$500.16.**
   $133.66 + $51.98 + $56.80 = $242.44 total deductions
   $742.60 − $242.44 = $500.16 net pay

5. Nick's net pay was **$19,183.98.**
   $5183.46 + $1439.85 + $2209.71 + $780.00 = $9613.02 total deductions
   $28,797 − $9,613.02 = $19,183.98 net pay

EXERCISE 34b (PAGE 188)

1. Leslie has **$296.75** left for other expenses.
   $225.00 + $94.50 + $45.00
   + $175.00 = $539.50 total expenses
   $836.25 − $539.50 = $296.75
   remaining

2. Howard has **$340.18** left after he pays his regular bills each month.
   $385.75 + $62.40 + $225.00
   + 136.67 = $809.82 total expenses
   $1150.00 − $809.82 = $340.18
   remaining

3. Tim saves **$215.20** each month.
   $384.50 + $74.00 + $175.00
   + $124.30 + $125.00 = $882.80 total expenses
   $1098 − $882.80 = $215.20
   remaining

4. Consuelo has **$5** left each month after she pays her regular expenses.
   $230 + $45 + $50 + $270 + $325
   = $920 total expenses
   $925 − $920 = $5 remaining

5. Terry had **$5087** left after paying his regular expenses.
   $4488 + $660 + $2220 + $2497
   = $9865 total expenses
   $14,952 − $9,865 = $5,087 remaining

6. Dan had **$1111.44** left for the rest of his expenses.
   $560.00 + $86.45 + $32.66
   = $679.11 total expenses
   $1790.55 − $679.11 = $1111.44
   remaining

7. Ingrid and Jake have **$12,025** left in their budget.
   $6000 + $1020 + $322 = $7342 total expenses
   $19,367 − $7342 = $12,025
   remaining

8. Sylvia has **$1358** after she pays her expenses.
   $480 + $82 + $250 + $82 + $246
   = $1140 total expenses
   $2498 − $1140 = $1358 remaining

EXERCISE 34c (PAGE 190)

1. Package (**2**) costs less per oz.
   (1) $2.25 ÷ 75 oz = $.03 per oz
   (2) $1.50 ÷ 60 oz = $.025 per oz

2. Container (**1**) costs less per fl oz.
   (1) $1.80 ÷ 120 fl oz = $.015 per fl oz
   (2) $1.02 ÷ 60 fl oz = $.017 per fl oz

3. Bottle (**1**) costs less per qt.
   (1) $.90 ÷ 1.5 qt = $.60 per qt
   (2) $1.00 ÷ 1.6 qt = $.625 per qt

4. Package (**1**) costs less per oz.
   (1) $1.98 ÷ 18 oz = $.11 per oz
   (2) $1.56 ÷ 13 oz = $.12 per oz

5. Box (**1**) costs less per lb.
   (1) $3.42 ÷ 4.5 lb = $.76 per lb
   (2) $3.00 ÷ 3.75 lb = $.80 per lb

6. Loaf (**2**) costs less per lb.
   (1) $1.29 ÷ 1.4 lb = $.921 per lb
   (2) $1.49 ÷ 2 lb = $.745 per lb

7. Piece (**1**) costs less per lb.
   (1) $2.56 ÷ .75 lb = $3.413 per lb
   (2) $1.44 ÷ .4 lb = $3.60 per lb

8. Turkey (**2**) costs less per pound.
   (1) $17.44 ÷ 16 lb = $1.09 per lb
   (2) $11.88 ÷ 12 lb = $.99 per lb

9. Can (**1**) costs less per ounce.
   (1) $1.22 ÷ 10 oz = $.122 per oz
   (2) $1.04 ÷ 8 oz = $.13 per oz

10. Can (**2**) costs less per ounce.
    (1) $1.39 ÷ 6 oz = $.232 per oz
    (2) $2.36 ÷ 12 oz = $.197 per oz

EXERCISE 34d (PAGE 191)

1. Janna runs **5024 yd**
   Circumference = 3.14 × 200 yd
   Circumference = 628 yd
   628 yd × 8 = 5024 yd

2. It will cost **$244.64** to resurface the floor.
   Area = 8 m × 5.5 m
   Area = 44 sq m
   $5.56 × 44 sq m = $244.64

3. It cost Kevin **$12,500** to fence the field.
   To find the perimeter, add the lengths of all four sides:
   .5 km + .75 km + .5 km + .75 km
   = 2.5 km
   $5000 × 2.5 km = $12,500

4. It will cost **$168.75** to cover the surface of the pool.
   Area = 25 ft × 15 ft
   Area = 375 sq ft
   $.45 × 375 sq ft = $168.75

5. The total weight of the tarpaulin is **4480** lb.
   Area = 40 yd × 40 yd
   Area = 1600 sq yd
   2.8 lb × 1600 sq yd = 4480 lb

6. Raymond will use **7.5 gal** of paint on the fence.
   Area = 30 yd × 5 yd
   Area = 150 sq yd
   150 sq yd ÷ 20 sq yd = 7.5 gal

7. It will cost **$1128** to carpet the room.
   Area = 12 yd × 4 yd
   Area = 48 sq yd
   $23.50 × 48 sq yd = $1128

8. The total cost for the court was **$7680**.
   Area = 75 ft × 32 ft
   Area = 2400 sq ft
   $3.20 × 2400 sq ft = $7680

DECIMALS REVIEW (page 193)

Part A

1. **55.6**	2. **31.08**	3. **.0126**
4. **.09**	5. **.0651**	6. **.775**
7. **16**	8. **8.411**	9. **.32**
10. **91.5**	11. **.108**	12. **47.9**
13. **9.3039**	14. **.775**	15. **.2301**
16. **7.4**	17. **.58**	18. **65.642**
19. **.264**	20. **32.26**	21. **40.294**
22. **.0014**	23. **72,390.5**	24. **61.98**
25. **60.671**	26. **1.085**	27. **40,032.748**
28. **3135.702**		29. **.175**
30. **753.76**		31. **851.7059**
32. **2160**		33. **332.88**

Part B

1. (a) **.06** (b) **.024** (c) **4.25** (d) **.055** (e) **30.3**
2. **4.34, 4.35, 4.53, 4.54**
3. **7.99, 7.42, 7.28, 7.038**
4. (a) **32.46** (b) **.4** (c) **73.00** (d) **$82.16** (e) **.6**
5. The meat costs **$7.85**.
   (2.25 lb × $3.49 = $7.8525)
6. The unit cost for the candy is **$.25 per oz.** ($1.99 ÷ 8 oz = $.24875)
7. The average is **20.582**. (8.5 + .98 + 84 + 7.5 + 1.93 = 102.91; 102.91 ÷ 5 = 20.582)
8. The perimeter of the square is **4.8 cm.** (1.2 cm × 4 = 4.8 cm)
9. The area of the rectangle is **35.75 sq km.** (5.5 km × 6.5 km = 35.75 sq km)
10. The perimeter of the triangle is **12 m.** (3 m + 4 m + 5 m = 12 m)
    The area of the triangle is **6 sq m.** (4 m × 3 m = 12 sq m; 12 sq m ÷ 2 = 6 sq m)
11. The circumference of the circle is **19.468 cm.** (The diameter = 6.2 cm; 3.14 × 6.2 cm = 19.468 cm)
    The area of the circle is **30.1754 sq cm.** (3.14 × 3.1 cm × 3.1 cm = 30.1754 sq cm)
12. Jerry's take-home pay was **$295.36**. ($73.91 + $31.41 + $5.63 + $4.44 = $115.39; $410.75 − $115.39 = $295.36)

13. After he pays his regular bills each month, Ali has **$476.85** left. ($385.75 + $62.40 + $225 = $673.15; $1150 − $673.15 = $476.85)
14. Brand (2) costs less per ounce. ($1.80 ÷ 24 = $.075; $1.40 ÷ 20 = $.07)
15. Herb runs **251.2 yd** when he goes around the track twice. (The diameter is 40 yards. 3.14 × 40 yd = 125.6 yd; 125.6 yd × 2 = 251.2 yd)

GED PRACTICE 2 (page 195)

1. **(2)** $265.50 + $434 + $376.50 + $282 = $1358.00
2. **(3)** $189 − $62.50 = $126.50
3. **(4)** 15 × .023 cm = .345 cm
4. **(2)** $44.59 ÷ 3 = $14.863, or $14.86 rounded to the nearest cent
5. **(4)** $16.98 ÷ $.849 = 20 gal
6. **(2)** 4 hr + 3 hr + 2 hr + 6 hr + 4 hr = 19 hr;
   19 hr × $6.35 = $120.65
7. **(1)** When you change 2.2 L to .0022 units of measure, you move the decimal point 3 places to the left. On the table of metric relationships (page 169), the column three to the left of liters is the kiloliter column.
8. **(2)** .1 in. + 1.1 in. + 2 in. + 9 in. + .6 in. + .2 in. + .9 = 5.8 in.;
   5.8 in. ÷ 7 = .8285 in., or .829 in. rounded
9. **(2)** 20.5 m + 30 m + 20.5 m + 30 m = 101 m
10. **(3)** 2.8 cm × 3 = 8.4 cm
11. **(5)** 12.5 m × 8.2 m = 102.5 sq m
12. **(3)** Because the radius is 6.5 cm, the diameter is 13 cm.
    Circumference = 3.14 × diameter
    Circumference = 3.14 × 13 cm
    Circumference = 40.82 cm
13. **(4)** Area = 3.14 × radius × radius
    Area = 3.14 × 3.5 ft × 3.5 ft
    Area = 38.465 sq ft
14. **(5)** Cost = number × rate
    Cost = 7 lb × $.79
    Cost = $5.53
15. **(1)** Unit cost = price ÷ unit
    Unit cost = $1.20 ÷ 15 oz
    Unit cost = $.08 per oz
16. **(2)** $126.58 + $47.47 + $31.60 + $11.75 = $217.40;
    $632.90 − $217.40 = $415.50

17. **(3)** $423.75 + $52.30 + $46 = $522.05;
$1000 − $522.05 = $477.95

18. **(3)** $1.60 ÷ 15 oz = $.107 per oz;
$1.65 ÷ 16 oz = $.103 per oz;
$2.05 ÷ 20 oz = $.1025 per oz;
$2.50 ÷ 24 oz = $.104 per oz;
$4.00 ÷ 36 oz = $.11 per oz

19. **(2)** $28.75 ÷ 20 = $1.4375 per trip, or
$1.44 rounded to the nearest cent;
$1.55 − $1.44 = $.11

20. **(2)** 10 m × 2 m = 20 sq m;
20 sq m ÷ 10 = 2 L

POSTTEST (page 199)

1. **(3)** 5039 ft × 6 = 30,234 ft
30,234 ft ÷ 5280 ft = 5.726 mi, or
rounded to the nearest hundredth,
5.73 mi

2. **(4)** 1 lb × 24 = 24 lb
9 oz × 24 = 216 oz
216 oz ÷ 16 oz = 13 lb 8 oz
13 lb 8 oz + 24 lb = 37 lb 8 oz

3. **(4)** 7 hr × 7 da = 49 hr
45 min × 7 da = 315 min
315 min ÷ 60 min = 5 hr 15 min
5 hr 15 min + 49 hr = 54 hr 15 min

4. **(2)** 6 T × 2000 lb = 12,000 lb
12,000 lb + 250 lb = 12,250 lb
12,250 lb ÷ 7 = 1750 lb

5. **(2)** 75 m + 100 m + 125 m = 300 m

6. **(2)** Area $= \dfrac{\text{base} \times \text{height}}{2}$
Area $= \dfrac{8\,\text{cm} \times 6\,\text{cm}}{2}$
Area = 24 sq cm

7. **(3)** 32 ft − 12 ft = 20 ft
27 ft − 12 ft = 15 ft
20 ft + 15 ft + 32 ft + 27 ft = 94 ft

8. **(4)** 32 ft × 27 ft = 864 sq ft
12 ft × 12 ft = 144 sq ft
864 sq ft − 144 sq ft = 720 sq ft

9. **(3)** 2454 tickets ÷ 7 volunteers = 409
tickets per volunteer

10. **(3)** 23 lines × 80 characters = 1840
characters

11. **(3)** 287 mi × 34 = 9758 mi

12. **(5)** 30,460 fans × 18 games = 548,280
fans

13. **(2)** 2478 employees ÷ 32 seats = 87
classes

14. **(3)** 9000 tickets − 783 tickets = 8217
tickets

15. **(4)** 7973 boys + 8638 girls = 16,611
children
16,611 children ÷ 6780 families =
2.45 children per family

16. **(3)** 750 sq ft × 2 floors = 1500 sq ft
1500 sq ft ÷ 3 = 500 sq ft

17. **(3)** 3 commercials × 8 hr = 24 hr
$15,000,000 ÷ 24 hr = $625,000
per hour

18. **(3)** Add zeros to compare the decimals:
.250, .500, .125, .375, .750.

19. **(5)** .75 hr × 12 km per hr = 9 km
9 km × 5 da = 45 km, or 45,000 m

20. **(3)** 15.6 mpg + 22.4 mpg + 8.1 mpg +
39.8 mpg + 19.3 mpg = 105.2 mpg
105.2 ÷ 5 = 21.04 mpg

21. **(4)** 34 m × 4 = 136 m

22. **(1)** 6 ft × 8 ft = 48 sq ft
Area = 3.14 × radius × radius
Area = 3.14 × 3 ft × 3 ft
Area = 28.26 sq ft
One piece of plywood covers an area
greater than the area of the pool.

23. **(3)** Circumference = 3.14 × diameter
Circumference = 3.14 × 12 ft
Circumference = 37.68 sq ft, or
rounded to the nearest foot, 38 ft

24. **(2)** Area = 3.14 × radius × radius
Area = 3.14 × 6 cm × 6 cm
Area = 113.04 sq cm, or rounded to
the nearest hundredth, 1.13 sq m

25. **(1)** $12.85 × 1.4 lb = $17.99
$20 − $17.99 = $2.01

26. **(2)** $246.35 × 4 = $985.40
$475 + $60.25 + $48.80 = $584.05
$985.40 − $584.05 = $401.35

27. **(4)** $ .87 ÷ 6 oz = $.145 per oz
$1.70 ÷ 12 oz = $.142 per oz
$2.24 ÷ 15 oz = $.149 per oz
$3.29 ÷ 24 oz = $.137 per oz
$3.50 ÷ 24 oz = $.146 per oz

28. **(5)** 5 m × 5 m = 25 sq m
5 m × 15 m = 75 sq m
25 sq m + 75 sq m = 100 sq m
100 sq m × $25 = $2500